Coagulation Kinetics and Structure Formation

Coagulation Kinetics
and Structure Formation

Hans Sonntag and Klaus Strenge

Academy of Sciences of the GDR
Berlin, GDR

English Language Scientific Editor:

B. Vincent

University of Bristol
England

PLENUM PRESS · NEW YORK AND LONDON

Library of Congress Catalog Card Number: 85-063436

The Authors are Hans Sonntag (Chapter 1—3)
and Klaus Strenge (Chapter 4)

Plenum Press, New York
A Division of Plenum Publishing Corporation
233 Spring Street, New York, N.Y. 10013

ISBN 0-306-42298-0

Published in coedition with VEB Deutscher Verlag der Wissenschaften, Berlin/GDR

Printed in the German Democratic Republic

Preface

Colloidal dispersions play a very important role in nature, industry, and daily life. Sometimes, long-term stability is observed or desired as in ferrofluids (composed of very small magnetic particles with radii of ≤ 10 nm), which must be stable even in external fields.

On the other hand, only short-term stable dispersions may be necessary during actual processing operations, for example, dispersions of magnetite particles during tape manufacture.

The stability of dispersions and many of their physical properties are related to the interaction between the particles in the dispersion medium, which may contain surfactants or macromolecular species.

If the net interparticle interaction forces are attractive, then aggregation may occur. Two general types of aggregation behavior may be distinguished: coagulation and flocculation. These two terms are frequently used synonymously but IUPAC has recommended the following definitions:

Coagulation implies formation of compact aggregates, leading to the macroscopic separation.

Flocculation implies the formation of a loose or open network, floc, which may or may not separate macroscopically.

Flocculation brought about by the simultaneous coadsorption of polymer molecules on two (or more) particles is referred to as bridging flocculation.

If coagulation results in the merging of two particles into one, as may occur with liquid droplets in emulsions, this process is referred to as coalescence.

H. Sonntag
K. Strenge

Contents

List of Symbols

a	Particle radius; lattice constant
A	Hamaker constant
c	Concentration
C_P	Capacity of a plate condensor
C_δ	Capacity of the Stern layer
d	Distance of closest approach
D	Diffusion coefficient; particle diameter
D_1	Diffusion coefficient of a single particle
D_{11}	Relative diffusion coefficient of two single particles ($D_{11} = 2D_1$)
E	Field strength; Young modulus; rate of energy dissipation
f	Friction coefficient; focal length
F	Force
g	Polarizability per unit volume; gravitation constant
G	Free energy; elastic modulus
h	Height of sediment; penetration depth; Planck constant
H	Magnetic field strength
i	Number of segments
i_0	Exchange current density
I	Diffusion flux; transmitted light intensity
I_0	Incident light intensity
k	Reaction (coagulation) constant ($k = 2k_S$)
k_B	Boltzmann constant ($1.3805 \, \text{J} \cdot \text{K}^{-1}$)
k_S	Smoluchowski irreversible coagulation rate constant ($6.05 \cdot 10^{-12} \, \text{cm}^3 \cdot \text{s}^{-1}$ at 298 K)
K	Structure "compression" modulus
l	Segment length
L	Length of particle chains
M	mass; refractive index relative to the medium
M	Magnetic induction
n	Number of bonds
n_i	Number of ions of kind i per unit volume
N_K	Coordination number
P	Shear stress
P_c	Probability for cluster formation

P_y	Yield stress
q	Cross section
r	Center-to-center distance; axial distances in ultracentifugation
$\langle r^2 \rangle^{1/2}$	Root mean square end-to-end distance
R	Radius of interaction
s	Dimensionless distance $\left(s = \dfrac{r}{a} \right)$
S	Entropy
t	Time
T	Temperature (Kelvin)
T_{ag}	Coagulation time [defined by equ. (3.9)]
T_{de}	Disaggregation time [defined by equ. (3.31)]
u	Dimensionless distance $\left(u = \dfrac{r - 2a}{a} \right)$
U	Internal energy
v	Velocity; particle volume; interaction volume
V	Interaction energy
V_A	van der Waals energy
V_B	Born repulsion energy
V_{el}	Double layer repulsion
V_{ster}	Steric hindrance of adsorbed macromolecules (or surfactants)
W	Stability ratio
x_δ	Minimum distance of ions from the interface
z	Particle concentration
z_i	Valence of ions
α	Solvent parameter
α_{coll}	Collision probability ($\alpha = 8\pi D_1 R$)
α_{ef}	Collision efficiency ($\alpha_{ef} = 1/W$)
$\Delta\alpha$	Anisotropy in the electric polarizability
β	Disaggregation probability London constant
γ	Shear deformation
$\dot{\gamma}$	Rate of shear
δ	Distance of the Stern layer from the interface, thickness of plates, thickness of adsorbed layers
δ_{ij}	Kronecker symbol, $\delta_{ij} = \begin{cases} 1, & i = j \\ 0, & i \neq j \end{cases}$
ε	Dielectric constant
ε_S	Dielectric constant of the Stern layer
η	Viscosity; dipole moment
$\varkappa^{-1}$	Debye parameter
λ	Wavelength
λ_L	Characteristic (London) wavelength

μ	Dipole moment
ξ	Decay length
Π	Pressure
ϱ	Distance between atoms in the solid state
$\varrho(x)$	Volume charge density at distance x from the surface
σ_0	Surface charge density
σ_s	Charge of the Stern layer
τ	Relaxation time; turbidity
φ	Volume fraction
χ	Flory-Huggins parameter
ψ_0	Surface potential
ψ_δ	Stern potential
ω	Angular velocity

Introduction

The properties of colloidal dispersions are dependent on the particle (aggregate) size distribution, on the particle (aggregate) shape, and on the net interaction forces between the particles.

Because of their small size, colloidal particles undergo Brownian motion just as molecules do. This results in collisions between the particles in colloidal dispersions.

The stability and other properties of colloidal dispersions depend on whether such collisions lead to aggregation. If all collisions were ineffective in this regard then the dispersion would be stable indefinitely. Collisions between particles may be caused not only by thermal motion but also by gravitational forces and by convective diffusion. However, our discussion will mainly be confined to Brownian coagulation.

Several types of interactions may occur as two particles approach one another: (1) electrostatic repulsion resulting from the overlapping of their electrical double layers, (2) van der Waals attraction (always present between molecules and particles), or (3) steric hindrance resulting from adsorbed molecules of solvent, surfactants, or macromolecules. The net superposition of all these interactions determines the energy change associated with aggregation and hence the stability.

The demands with respect to the stability of colloids are quite different depending on the application. Short term stability of a colloidal dispersion may be preserved by stirring or shaking, even when that dispersion is thermodynamically unstable.

In many other cases long-term stability is required, for example for polymer latexes or pigmented paints. Such dispersions should be stable or should only aggregate reversibly. During the process of oil drilling dispersions with low viscosity (with no aggregation) are required. However, when the drilling process is stopped, a structuring of the dispersion is required in order to reduce sedimentation of the cuttings; for this step, weak, reversible aggregation is required.

Coagulation kinetics is classified as being "rapid" or "fast" if every collision is effective and "slow" if not. The latter case arises if either an energy barrier, comparable to the thermal energy ($k_B T$) of the particles, or a low shallow energy minimum, also comparable to $k_B T$, is present, where k_B is the Boltzmann constant and T is temperature.

It is necessary in interpreting coagulation kinetics to consider the mutual diffusion of the particles and the interaction between them. One of the principal concerns of colloid science is to accurately describe the nature of the interaction forces between the particles. However, it is not the aim of this book to give a detailed description of the different kinds of interaction forces. Insofar as it is necessary for the understanding of coagulation kinetics and the formation of structures, a short review of this topic is given in Chapter 1.

1.   Interaction between Colloidal Particles

Colloidal particles in air or liquids are mostly charged, owing either to the adsorption of charge carriers (ions or electrons) or to the dissociation of polar groups at the interface. In a colloidal dispersion in water all the particles of the same kind will have the same surface charge (and surface potential), if the surface charge is due to dissociation of polar groups or adsorption of potential determining ions. In polymer latexes, where the charge arises from hydrolysis of polar groups at the interface, the surface charge density of different particles may vary. Nevertheless, the electrostatic interaction is still a repulsive one. However, in nonpolar liquids or in air, dispersed particles can even have charges of different signs.

We shall confine our discussions to double layer forces in aqueous media. Van der Waals forces exist between all particles and, because these are long range, they play an important role in the total interaction. Additionally so-called repulsion forces arise from the interpenetration of any adsorbed layers of either low-molecular-weight species or macromolecules. Similarly adsorbed water molecules give rise to the so-called "hydration forces".

The consideration of colloid stability in terms of the superposition of the double layer repulsion and van der Waals attraction was developed independently by Derjaguin and Landau and by Verwey and Overbeek (DLVO theory). A survey of surface forces was recently published by Derjaguin et al. [41].

1.1.   Electrical Double Layer Interaction

Independently of the mode of charging, the surface charge of the particles is compensated by the charge of the counterions in the double layer. The structure of the double layer is shown in Figure 1.1; that part of the double layer extending from the surface of the particle to the minimum distance of the hydrated counterions is called the Stern layer. Within the Stern layer partly desolvated ions adsorbed by electrostatic attraction or specifically adsorbed ions may be present. These ions form the inner Helmholtz plane. The outer Helmholtz plane consists of the first layer of hydrated counterions also adsorbed by electrostatic attraction. (In our consideration we will not distinguish between these two layers; we shall simply refer to the "Stern layer".)

The remainder of the counterions, together with the coions, are distributed in the diffuse double layer. The ion distribution within this layer is described by the Gouy-Chapman theory.

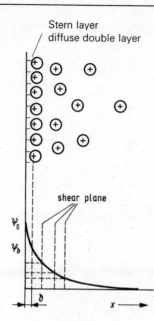

Figure 1.1
Scheme of the double layer.

Three basic equations describe the electrical double layer: the Poisson equation,

$$\text{div}\,[\varepsilon(x)\,\text{grad}\,\psi(x)] = -4\pi\varrho(x) \tag{1.1}$$

where ε is the dielectric constant of the solution, ψ the potential, ϱ the volume charge density, and x the distance from the surface; the Boltzmann equation,

$$n_i(x) = n_i(\infty)\exp-\frac{W_i(x)}{k_B T} \tag{1.2}$$

where $n_i(\infty)$ is the concentration of the ions of kind i in bulk solution and $W_i(x)$ the energy required to bring an ion i from the bulk ($x = \infty$) to distance x from the interface; and the equation

$$\varrho(x) = \sum z_i e n_i(x) \tag{1.3}$$

where $\varrho(x)$ is the volume density of charge, and z_i the valence of the ions. The Poisson-Boltzmann equation results if we assume ε is a constant at any distance from the interface, i.e.,

$$\frac{d^2\psi(x)}{dx^2} = -\frac{4\pi e}{\varepsilon}\sum z_i n_i(\infty)\exp-\frac{z_i e\psi(x)}{k_B T} \tag{1.4}$$

This equation may be integrated with the boundary conditions

$$x \to \infty, \quad \psi = 0, \quad \text{and} \quad \frac{d\psi}{dx} = 0 \tag{1.5}$$

In the Gouy-Chapman approach, the following assumptions are made:

1. plane double layer,
2. dielectric constant independent of x,
3. $W_i(x)$ only dependent on the Coulomb interaction, i.e., $W_i(x) = z_i e \psi(x)$.

This leads to the following relationship:

$$\varkappa x = \ln \frac{\tanh\left(\dfrac{z_i e \psi_0}{4 k_B T}\right)}{\tanh\left(\dfrac{z_i e \psi(x)}{4 k_B T}\right)}. \tag{1.6}$$

where $\varkappa^{-1}$, the Debye length, is defined by:

$$\varkappa = \left[\frac{4\pi e^2 \sum n_i(\infty) z_i^2}{\varepsilon k_B T}\right]^{1/2} \tag{1.7}$$

For the surface charge density we obtain:

$$\sigma_0 = -\frac{\varepsilon}{4\pi}\left(\frac{d(\psi)}{dx}\right)_{x=0} = \left[\frac{\varepsilon k_B T n_i(\infty)}{2\pi}\right]^{1/2} \sinh\left(\frac{z_i e \psi_0}{2 k_B T}\right) \tag{1.8}$$

According to the Stern theory the surface charge density equals the sum of the charge density of the Stern layer σ_S and that of the diffuse double layer σ_δ:

$$\sigma_0 = \sigma_S + \sigma_\delta \tag{1.9}$$

For the total capacity of the double layer, we obtain:

$$C = \frac{C_p \cdot C_\delta}{C_p + C_\delta} \tag{1.10}$$

where C_p is the capacity of the plate capacitor and C_δ the capacity of the diffuse part of the double layer. The capacity of the plate capacitor is almost independent of the electrolyte concentration and is given by

$$C_p = \frac{\sigma_0}{\psi_0 - \psi_\delta} = \frac{\varepsilon_\varrho}{4\pi x_\delta} \tag{1.11}$$

x_δ is the minimum distance of adsorbed ions from the interface.

The capacity of the diffuse part of the double layer equals

$$C_\delta = \frac{\sigma_\delta}{\psi_\delta} \tag{1.12}$$

and C_δ is proportional to the ion concentration $c_\pm$; therefore the Stern potentials also depend on the ion concentration. For σ_δ we derive

$$\sigma_\delta = -\left(\frac{\varepsilon k_B T c_\pm}{2\pi}\right)^{1/2} \sinh \frac{z e \psi_\delta}{2 k_B T} \tag{1.13}$$

The Poisson-Boltzmann equation for spherical particles, with the Gouy-Chapman approximations, has the following form:

$$\frac{1}{r^2} \cdot \frac{d}{dr} \cdot \left(r^2 \frac{d\psi(r)}{dr} \right) = \frac{4\pi e}{\varepsilon} \sum n_i(\infty)\, z_i \exp - \frac{z_i e \psi(x)}{k_B T} \tag{1.14}$$

where r is the center-to-center distance. The Poisson-Boltzmann equation had already been solved by Debye and Hückel [2] for weakly charged particles. This was obtained by a series expansion with truncation after the second term

$$\psi(r) = \psi_0 \frac{a}{r} \exp\left[-\varkappa(r-a)\right] \tag{1.15}$$

and

$$\sigma_0 = \frac{\varepsilon \psi_0}{4\pi a} (1 + \varkappa a) \tag{1.16}$$

The most detailed double layer calculations have been published by Loeb, Wiersema, and Overbeek in the form of tables [1]. They suggested the following empirical equation for the relationship between surface charge density and surface potential, based on numerical calculations:

$$\sigma_0 = \frac{\varepsilon k_B T \varkappa}{2\pi z_i e} \left[\sinh\left(\frac{e z_i(\psi_\delta)}{2k_B T}\right) + \frac{2}{\varkappa a} \tanh\left(\frac{e z \psi_\delta}{4k_B T}\right) \right] \tag{1.17}$$

Equation (1.17) gives fairly exact values for the condition $\varkappa a > 1$.

A fairly good approximation for the potential distribution within the equilibrium double layer around a spherical, nonconducting particle was developed by Martynov [3]:

$$\psi(\varkappa r) = \frac{1}{\sqrt{3B}} \text{ arc tanh} \left[\sqrt{3B}\, \xi\, \frac{\exp\left[-\varkappa(r-a)\right]}{\dfrac{r}{a}} \right] \tag{1.18}$$

with

$$\xi = \frac{1}{\sqrt{3B}} \tanh\left(\sqrt{3B}\psi_0\right) \tag{1.19}$$

and

$$B(\varkappa a) = \frac{a^2}{6} \left[2 \exp(4\varkappa a) - \exp(2\varkappa a) \right] E_i(2\varkappa a) \tag{1.20}$$

$$E_i(\varkappa a) = \int_a^\infty \frac{\exp - k_B T}{t}\, dt \tag{1.21}$$

The function $B(\varkappa a)$ is tabulated in reference [3]. Equation (1.18) has been shown to be accurate under the conditions $\dfrac{ze\psi_0}{k_B T} \leq 5$ and at any $\varkappa a$, or for any ψ at $\varkappa a > 3$.

Duchin and co-workers [4—6] have developed a method of successive approximations for the solution of the Poisson-Boltzmann equation for spherical and cylindrical particles where $(\varkappa a)^{-1}$ is small. The relative errors of the calculated values compared with the numerical calculations in [1] are tabulated in reference [7].

Electrostatic repulsion forces occur if the double layers of two approaching particles penetrate each other. The repulsion in fact results from the combination of an electrostatic and an osmotic interaction term. This kind of calculation was first made by Derjaguin and Landau [9—11] and, independently, by Verwey and Overbeek [12].

If two particles approach one another as a result of their Brownian motion (see chapter 3) then electrostatic repulsion will occur if their diffuse double layers overlap. If we assume full equilibrium relaxation of the double layers during their approach, then the electrostatic interaction must be calculated with the condition of constant potential. But if the time of the particle encounter is short compared to the time of the double layer relaxation, then electrostatic interaction should be calculated at constant charge. We will come back to this point in section 1.4.

To calculate the electrostatic interaction between spherical particles the nonlinear Poisson-Boltzmann equation (1.14) has to be solved. Until now no closed form solution for this equation has been obtained. With certain assumptions, simple approximate expressions are available. In particular the potential energy of interaction for large particles (with the assumption $\varkappa a \gg 1$) may be calculated using the Derjaguin approximation [8, 13]. The repulsive energy is calculated for a series of infinitesimal parallel rings. The total potential energy is then found by integrating over the whole surface of the interacting spherical particles. This procedure is valid for high electrolyte concentration and/or large particles, and for Stern potentials $\psi_\delta < 25$ mV. Derjaguin calculated, with the Debye-Hückel approximation for the potential distribution [equ. (1.15)], an equation for the electrostatic interaction energy:

$$V_{el} = \frac{\varepsilon a z \psi_\delta^2}{2} \ln \left[1 + \exp \left\{ -\varkappa(r - 2a) \right\} \right] \tag{1.22}$$

In many cases this equation can be applied with sufficient accuracy ($\pm 10\%$).

A better approach for calculations of the electrostatic interaction was published by McCartney and Levine [14], also valid for the condition $\varkappa a \geqq 5$. They obtained the following equation:

$$V_{el} = \frac{\varepsilon a z \psi_\delta^2 (r - 2a)}{r} \ln \left[1 + \frac{a}{r - a} \exp \left\{ -\varkappa(r - 2a) \right\} \right] \tag{1.23}$$

Recently an extended Hogg-Healy-Fuerstenau (HHF) equation[1]) [17] was derived for the interaction of unequal spheres [15, 16]. This may be expressed for spheres of

---

[1]) The HHF equation is the equivalent equation to the Debye-Hückel equation (1.22) for unequal spheres.

equal size in the following form:

$$V_{el} = \frac{1}{2}\,\varepsilon a\psi_\delta^2\,\left\{\ln\left(1 + \exp\left\{-\varkappa(r - 2a)\right\}\right) - \frac{1}{\varkappa(r - 2a)}\right.$$

$$\times\left[\frac{1}{6}\,\varkappa(r - 2a)\ln\left\{1 + \exp\left[-\varkappa(r - 2a)\right]\right\}\right.$$

$$+ \frac{\varkappa(r - 2a)\exp\left\{-\varkappa(r - 2a)\right\}}{3\left[1 + \exp\left\{-\varkappa(r - 2a)\right\}\right]^2} - \left.\frac{\exp\left\{-\varkappa(r - 2a)\right\}}{3(1 + \exp)\left\{-\varkappa(r - 2a)\right\}}\right]$$

$$\left.+ \frac{1}{3}\sum_{n=1}^{\infty}\frac{(-1)^{n-1}}{n^2}\exp\left\{-n\varkappa(r - 2a)\right\}\right\} \tag{1.24}$$

The interaction energy under conditions of small $\varkappa a$ may be calculated using the linear approximation. This is described in detail in reference [12]. All the calculations for small $\varkappa a$ (section 1.4.) were made with the following equations. If constant potential is assumed during the particle encounters, we obtaine

$$|V_{el}|_{\psi_\delta = \text{const}} = \psi_\delta^2\varepsilon a\,\frac{\exp\left[-\varkappa(r - 2a)\right]}{\dfrac{r}{a}}$$

$$\times\frac{1 + \alpha}{1 + \dfrac{\exp\left[-\varkappa(r - 2a)\right]}{2\varkappa r}\left[1 - \exp\left(-2\varkappa a\right)\right](1 + \alpha)} \tag{1.25}$$

$$\alpha = \lambda_1\left(1 + \frac{1}{\varkappa r}\right) + \lambda_2\left(1 + \frac{3}{\varkappa r} + \frac{3}{(\varkappa r)^2}\right) \tag{1.26}$$

λ_1 and λ_2 are defined by

$$\frac{1}{3}\,\lambda_1 + \frac{\exp\left[-\varkappa(r - 2a)\right]}{2\varkappa r}\left\{\frac{\varkappa a - 1}{\varkappa a + 1} + \exp\left(-2\varkappa a\right)\right\}$$

$$\times\left\{1 + \frac{1}{\varkappa r} + \lambda_1\left(1 + \frac{2}{\varkappa r} + \frac{2}{(\varkappa r)^2}\right) + \lambda_2\left(1 + \frac{4}{\varkappa r} + \frac{9}{(\varkappa r)^2} + \frac{9}{(\varkappa r)^3}\right)\right\} = 0$$

$$\tag{1.27}$$

and

$$\frac{1}{5}\,\lambda_2 + \frac{\exp\left[-\varkappa(r - 2a)\right]}{2\varkappa r}\left\{\frac{(\varkappa a)^2 - 3\varkappa a + 3}{(\varkappa a)^2 + 3\varkappa a + 3} - \exp\left(-2\varkappa a\right)\right\}$$

$$\times\left\{1 + \frac{3}{\varkappa r} + \frac{3}{(\varkappa r)^2} + \lambda_1\left(1 + \frac{4}{\varkappa r} + \frac{9}{(\varkappa r)^2} + \frac{9}{(\varkappa r)^3}\right)\right.$$

$$\left.+ \lambda_2\left(1 + \frac{6}{\varkappa r} + \frac{24}{(\varkappa r)^2} + \frac{54}{(\varkappa r)^3} + \frac{54}{(\varkappa r)^4}\right)\right\} = 0 \tag{1.28}$$

If constant charge is assumed during the encounter, we obtain

$$|V_{el}|_{\sigma_0 = const} = \psi_0^2 \varepsilon a \ \frac{\exp\left[-\varkappa(r - 2a)\right]}{\dfrac{r}{a}}$$

$$\times \frac{1 + \alpha}{1 - \dfrac{\exp\left[-\varkappa(r - 2a)\right]}{2\varkappa r} \left[\dfrac{\varkappa r - 1}{\varkappa r + 1} + \exp\left(-2\varkappa a\right)\right] \left[1 + \alpha\right]}$$

(1.29)

For the construction of the interaction energy distance curves in section 1.4., we used for $\varkappa a > 5$ equation (1.22) and (1.23) and for lower $\varkappa a$ (1.25) and (1.29).

1.2.   van der Waals Interaction Energy

The van der Waals energy is always an attraction energy between particles of the same kind and consists of three components: the London or dispersion energy; the Keesom or dipole orientation energy; and the Debye or induction energy. The dispersion energy in particular is of great importance in coagulation.

The dispersion energy is the result of charge fluctuations in the atoms associated with the motion of the electrons. These charge fluctuations produce a time dependent dipole moment. A phase difference in the fluctuating dipoles leads to mutual attraction. The van der Waals energy can be calculated in two different ways.

Hamaker derived the attraction energy between two particles by pairwise integration of the interatomic dispersion energy over the interior volume of the two interacting particles [18—20].

The Hamaker theory leads to the expression

$$V_A = - \int\limits_{v_1} \int\limits_{v_2} \frac{q^2 \beta}{s^6} \, dv_1 \, dv_2$$

(1.30)

with v_1, v_2 the volumes of two particles, dv_1, dv_2 the volume elements at distance s, q the number of atoms, and β the London constant

$$\beta = \frac{3}{2} h \ \frac{v_{01} v_{02}}{v_{01} + v_{02}} \alpha_{01} \alpha_{02}$$

(1.31)

with α_{01}, α_{02} the polarizability of two atoms, v_{01}, v_{02} their characteristic frequencies, and h the Planck constant.

The Lifschitz theory [21, 22] considers the bulk electrodynamic response of body 1 to all possible electrodynamic fluctuations in body 2 and vice versa. This theory calculates the van der Waals attraction energy from macroscopic properties and makes it possible to separate the van der Waals energy into contributions from three separate frequency regimes, each ascribed an average wavelength. Despite a number of shortcomings the Hamaker approach is often used because of its greater convenience and also because the results are a reasonable approximation, particularly for non-polar systems. After integration of equation (1.30) for spherical symmetry the following equation is obtained for the van der Waals interaction of two unequal spheres:

$$V_D = - \frac{A}{6} \left[\frac{2a_1 a_2}{d^2 + 2a_1 d + 2a_2 d} + \frac{2a_1 a_2}{d^2 + 2a_1 d + 2a_2 d + 4a_1 a_2} \right.$$
$$\left. + \ln \frac{d^2 + 2a_1 d + 2a_2 d}{d^2 + 2a_1 d + 2a_2 d + 4a_1 a_2} \right] \tag{1.32}$$

with $d = r - 2a$ the distance of closest approach and the Hamaker constant A

$$A = \beta \pi^2 q^2 \tag{1.33}$$

The approximate equation (1.32) may be further simplified when the particle radii are much larger than the interparticle distance; equation (1.34) is then applicable:

$$V_D = - \frac{Aa}{12d}; \qquad a \gg d \tag{1.34}$$

Because of the finite propagation velocity of the electromagnetic waves the phase shift of the fluctuating dipoles will differ from 180° for distances between two atoms which are of the same magnitude as the London wavelength λ_L. Consequently the dispersion forces will decrease more rapidly than predicted by equation (1.30). The correction factor for the London coefficient β in equation (1.31) is given by

$$f = \left(\frac{2\pi s}{\lambda_L} \right)$$

λ_L is the characteristic wavelength for the interaction, which is often assumed to be about 100 nm.

Overbeek [23] was not able to find a single expression for the retardation and therefore the Hamaker integration procedure becomes considerably more difficult for spheres than for plates. Overbeek introduced the following two equations:

$$f \left(\frac{2\pi s}{\lambda_L} \right) = 1.01 - 0.14 \frac{2\pi s}{\lambda_L} \quad \text{for} \quad 0 < \frac{2\pi s}{\lambda_L} < 3 \tag{1.35}$$

and

$$f \left(\frac{2\pi s}{\lambda_L} \right) = 2.45 \frac{\lambda_L}{2\pi s} - 2.04 \frac{\lambda_L^2}{4\pi^2 s^2} \quad \text{for} \quad 3 \leqq \frac{2\pi s}{\lambda_L} \tag{1.36}$$

Schenkel and Kitchener [24] found a single expression for the retarded van der Waals force between two atoms:

$$f \left(\frac{2\pi s}{\lambda_L} \right) = 2.45 \frac{\lambda_L}{2\pi s} - 2.17 \frac{\lambda_L^2}{4\pi^2 s^2} + 0.9 \frac{\lambda_L^3}{8\pi^3 s^3} \tag{1.37}$$

These three equations have frequently been applied in equation (1.30).

The usual basis for calculations of this type is the Casimir-Polder equation [26] in which all the contributions to the interaction between two neutral atoms are considered. Using these equations relationships for the van der Waals interaction have been arrived at by Clayfield, Lumb, and Mackey (CLM) [27]. Their expression is complicated but well suited to rapid evaluation by computer.

All the calculations described in section 1.4. were performed using this equation.

Recently Gregory [25] compared calculations based on these various approximate expressions with those based on the "exact" CLM equation. He found for particles with radii 1 μm (see Fig. 1.2) good agreement between the "exact" equation and the Kitchener approach [24]

$$V_A = \frac{Aa}{12(r - 2a)} \left[\frac{1}{1 + 11.12 \dfrac{r - 2a}{\lambda_L}} \right] \tag{1.38}$$

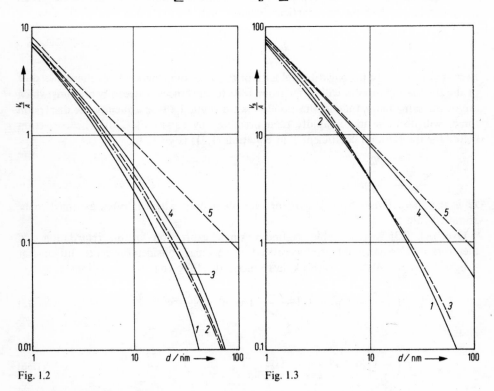

Fig. 1.2   Fig. 1.3

Figure 1.2
Dimensionless van der Waals energy—distance curves of equal spheres with $a = 1$ μm.
1, "exact retarded result" [27]; 2, Gregory equation (1.42); 3, Kitchener equation (1.39); 4, equation (1.32), 5 equation (1.35).

Figure 1.3
Dimensionless van der Waals energy—distance curves for equal spheres with $a = 100$ nm (cf. Fig. 1.2).

and also with the Gregory equation:

$$V_A = - \frac{Aa}{12(r - 2a)} \left[1 - \frac{b(r - 2a)}{\lambda_L} \left(1 + \frac{\lambda_L}{b(r - 2a)} \right) \right] \qquad (1.39)$$

The constant b was chosen to give the best fit with calculations based on the CLM equation. A value of $b = 5.32$ gave satisfactory results if the separation is smaller than the particle radius.

Equations (1.38) and (1.39) may also be used for calculations with unequal spheres if $\frac{a}{12}$ and $r - 2a$ is replaced by

$$\frac{\sqrt{a_1 a_2}}{12(r - a_1 - a_2)}$$

For particles at surface—surface distances large compared to the particle radius the agreement between the approximate equations and the full CLM equation was poor (Fig. 1.3). However, for small distances equations (1.32), (1.38), and (1.39) give good agreement.

All the equations described above were derived for smooth particles, but in some cases the particles may be rough or porous. For the electrostatic interaction between colloidal particles, this effect may be of no real significance, because the counterion distribution is smeared out by the Brownian motion of the counterions in the diffuse double layer around a rough particle.

On the other hand, the roughness of the interacting particles may greatly affect the van der Waals attraction. This point was considered recently by Czarneckij and Itšenskij [28].

If we assume that any surface irregularities are confined for the two particles within a shell of thickness B_1 and B_2 (see Fig. 1.4), then each particle consists of a solid core with constant density and a shell with a radial density distribution. Details of the analysis are given in reference [29]. Therefore the total van der Waals interaction is divided into four contributions, namely, the interaction between core *1* and core *2*, between shell *1* and shell *2*, between core *1* and shell *2*, and

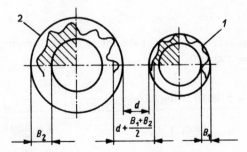

Figure 1.4
Interaction of two rough particles (schematically).

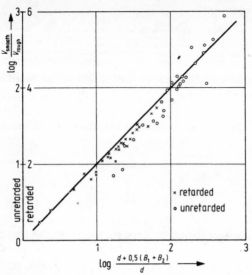

Figure 1.5

The relative attraction energy $\dfrac{V_{\text{smooth}}}{V_{\text{rough}}}$ as a function of the dimensionless distance $1 + 0.5 \left[\dfrac{(B_1 + B_2)}{d} \right]$ in logarithmic scale.

between core *2* and shell *1*. The necessary integration procedure was carried out both for the nonretarded and the retarded cases. The expressions obtained are rather lengthy and complicated. They may be obtained from the authors in the form of a FORTRAN subroutine listing.

The van der Waals energies for rough particles were compared with those for smooth particles calculated using the Gregory equation (1.39). The results are shown in Figure 1.5 for the retarded and nonretarded cases. It may be seen that the slope of the line representing the nonretarded case is approximately unity; therefore a correction factor for surface roughness can be introduced into the Gregory equation (1.39), i.e.,

$$\frac{d}{d + 0.5(B_1 + B_2)}$$

Therefore, the following equation is obtained for the interaction of rough (or porous) particles of different radii ($\lambda_L = 100$ nm):

$$V_A = -A \; \frac{a_1 a_2 \left[1 - 5.32\ 10^5 d \log \left(1 + \dfrac{1}{5.32\ 10^5 d} \right) \right]}{6d(a_1 + a_2)} \cdot \frac{d}{d + 0.5\,(B_1 + B_2)}$$

$$(1.40)$$

where d is the distance between the outermost peaks of the rough interfaces.

Effects resulting from surface roughness in the total interaction between particles may be of considerable influence on the reversibility of coagulation in the primary minimum. We will come back to this point in section 1.4.

It should be mentioned that the Hamaker theory has some limitations especially at short distances, for example, the neglect of the finite size of atoms, the introduction of a single wavelength λ_L, and pairwise summation, which neglected many-body effects. Moreover, at the interface there may be a layer of oriented permanent dipoles (for example, water molecules) or ions having a different van der Waals interaction from the actual molecules comprising the particles.

Therefore, it is very difficult to make predictions regarding the adhesion between coagulated particles. For example, the distance of closest approach between molecular smooth particles is about twice the radii of the atoms concerned. Very strong repulsion forces (Born forces) occur if two atoms (molecules) come into contact, that is if their electron clouds start to interpenetrate. From quantum mechanical considerations an exponential decay with distance may be assumed, i.e.,

$$V_B = \sim \exp\left(-\frac{v}{\xi}\right) \tag{1.41a}$$

with s the distance between atoms in the solid state and ξ the decay length. Very often the Born repulsion is approximated by the empirical equation

$$V_B = \frac{v}{s^{12}} \quad (v \text{ material constant}). \tag{1.41b}$$

From this it follows that calculation of the total attraction energy between particles in the primary minimum is not possible, if we have no information about surface inhomogeneities or adsorption layers. At the end of this chapter some typical Hamaker constants in water are tabulated (Table 1.1).

Table 1.1

Hamaker constants $A \cdot 10^{20}$ J for different substances in water

Substance	$A \cdot 10^{20}$ J
Polystyrene	0.35
Ionic crystals	0.3—4.8
Gold	10
Silver	1—4
Quartz	1.2
Oxides	1.7—4
Diamond	14

1.3.   Steric Interaction (Structural Forces) Arising from Adsorbed Molecules

Structural forces between approaching particles arise from the penetration of their adsorbed layers of solvent, surfactant, or macromolecules. A general review was recently published by Derjaguin et al. [41].

If one assumes only one monolayer of oriented adsorbed water molecules at each interface and neglects dipole—dipole and dispersion interactions between these two layers, then the primary minimum energy is reduced, simply because the distance of closest approach of the interfaces is now increased. From the collision diameter of water molecules (0.46 nm) the minimum distance of approach is 0.92 nm, leading to a reduction in the minimum energy of about three quarters.

The first experimental isotherms of structural forces were measured by Derjaguin et al. [42] at the water-quartz interface by measuring the thickness of water films as a function of the vapor pressure by ellipsometry [42]. The decay length ξ of the hydration forces was found to be 10 nm. That is a rather high value. This result was later confirmed by Pashley and Kitchener [43]. They found that both clean and hydroxylated quartz form substantially thicker films than would be expected from DLVO theory. Rabinovič et al. [44] measured interaction forces between smooth crossed quartz

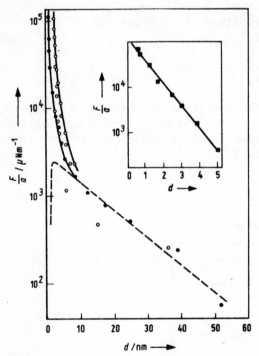

Figure 1.6

Hydration forces—distance curve according to Pashley [30]; dashed line: superposition of electrostatic repulsion and van der Waals attraction.

filaments and found a decay length of 7—8 nm. Many experiments about solvation forces were performed between crossed mica cylinders by Israelachvili [45] and Pashley and co-workers [30, 48, 49]. It was found that the repulsion force is due to the hydration of the counterions. The decay length of these forces is about 0.9 nm. A typical curve is shown in Figure 1.6.

From experimental pressure-distance isotherms the structural component of the inter-action energy can be calculated by means of equation (1.42):

$$V_{ster} = \int_{d_{min}}^{\infty} \Pi_{ster}(d) \, dd = K\xi \exp(-d/\xi) \tag{1.42}$$

with ξ the decay lenght, and K a preexponential constant (for crossed quartz filaments K was found to be 10^3 mN $\cdot$ m^{-2}) [41].

The counterion hydration mechanism ought to be valid for any coagulation system where the surface has both a high negative surface charge density and little or no possibility for hydrogen bonding. Monodisperse amphoteric polystyrene latices are a perfect example of such a system. Healy et al. [46] have found that amphoteric latices are stable at high pH one or two units above the isoelectric point (iep) even at 1 mol $\cdot$ dm^{-3} salt with K$^+$ and to a greater extent with Li$^+$ [46]. Ruckenstein and Schiby [47] postulated that the adsorbed ions generate a bilayer of charge which then generates a sufficiently strong polarization of the neighboring water molecules which propagates further into the liquid as a result of the interactions among the water molecules.

Influence of adsorbed surfactant molecules and macromolecules on the interaction

Surfactants or macromolecules are often needed to modify colloid stability. An increase in the stability of dispersed particles in water through adsorption of surfactants can only arise if the hydrophilic parts of these molecules are oriented towards the solution. In apolar solvents exactly the opposite orientation of the adsorbed molecules is demanded. We will confine discussion to aqeous dispersions.

If the adsorption is caused by the so called "hydrophobic interaction", then the polar groups are oriented towards the water phase. The origin of the hydrophobic interaction is the very strong interaction between water molecules compared to that between water molecules and organic groups like CH_3- or CH_2-. Therefore, the apolar parts of bipolar molecules are rejected from the water phase.

If the polar groups are charged then the electrostatic interaction is as described in section 1.1. and the van der Waals interaction as in section 1.2. If we ascribe the adsorbed layer its own Hamaker constant then the following equation is obtained from the attraction energy [31]:

$$V_A = -\frac{1}{12}\left[(A_0^{1/2} - A_2^{1/2})^2 \frac{a+\delta}{d} + (A_2^{1/2} - A_1^{1/2})^2 \frac{a}{d+2\delta}\right.$$

$$\left. + 4(A_0^{1/2} - A_2^{1/2})(A_2^{1/2} - A_1^{1/2}) \frac{a+\delta}{(d+\delta)\left(2+\dfrac{\delta}{a}\right)} \right] \tag{1.43}$$

δ is the thickness of the adsorbed layer, A_0 is the Hamaker constant of the solvent, A_1 is the Hamaker constant of the particle, A_2 is the Hamaker constant of the adsorbed layer, and $d = r - 2(a + \delta)$.

On the other hand adsorption of surfactants may occur with the polar groups towards the surface, for example, fatty acids at oxidic particles. In such cases a second layer of surfactant molecules may adsorb through hydrophobic interaction with the polar groups oriented outward. Such surfactant "double layers" are very important and very effective for example in stabilization of ferrofluids in water [32]. As with "bare" particles the stability is dependent on the superposition of the van der Waals attraction, electrostatic repulsion, and Born repulsion.

For particles with nonionic surfactants adsorbed by hydrophobic interaction a steric repulsion occurs. The magnitude of this repulsion energy increases rather steeply with increasing penetration of the adsorbed layers. On the other hand, because of the strong hydration of the ethoxy groups the van der Waals attraction becomes very low. Hence such dispersions are very stable.

The adsorption of neutral macromolecules differs from the adsorption of nonionic surfactants in several ways:

— Commercially available macromolecules generally have a broad size distribution.
— The conformation of macromolecules in the adsorbed state is different from the conformation in the solution. In solution mostly polymers have a random coil conformation, whereas in the adsorbed state some segments are adsorbed at the interface (in trains), while the rest is distributed in loops or tails.

Loops consist of segments only in contact with solvent molecules and they are bound at the interface at every side by trains. The tail is terminally bound to a train; the other part dangles into the solution.

Scheutjens and Fleer [39] have calculated distribution functions for mixtures of free and adsorbed polymer chains and solvent molecules. They determined the equilibrium segment concentration profile using an iterative procedure. They first determined the number of weighted conformations of adsorbed chains, subject to an initial choice of segment concentration profile, and hence the associated partition function. They then search for the minimum partition function situation.

Some results of their computations are shown in Figure 1.7. The tail length increases rapidly with chain length (molecular weight), but the loop size is only slightly dependent on the molecular weight. It follows that steric hindrance during particle encounters is mainly determined by long, dangling tails.

Adsorbed neutral macromolecules can influence the electrostatic interaction and the van der Waals interaction, as well as giving rise to the steric repulsion.

We have investigated the influence of polyvinyl alcohol on the surface charge density of silica [33]. In this case we did not find any changes in the surface charge density. This was attributed to the fact that the adsorption of PVA molecules is caused by hydrophobic interaction. However with most systems it is necessary to check whether the adsorption of macromolecules leads to changes in surface charge density or in the distribution of the counterions around the particles.

Adsorbed macromolecules will also influence the van der Waals interaction. In particular the layers near the interface with higher segment density may contribute significantly to the attraction energy. For the calculation equation (1.40) can be used if B_1 and B_2 include the segment density distribution within the adsorbed layer.

The Hamaker constant of the adsorbed layer A_2 has been shown by Vincent [34] to be given by

$$A_2 \simeq [(1 - \varPhi_{\mathrm{M}})\, A_{\mathrm{H_2O}}^{1/2} + \varPhi_{\mathrm{M}} A_{\mathrm{M}}^{1/2}]^2 \tag{1.44}$$

$\varPhi_{\mathrm{M}}$ is the volume fraction of the polymer phase with the Hamaker constant A_{M}.

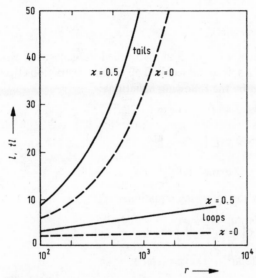

Figure 1.7
Average loop (*l*) and tail lengths (*tl*) for different solvent parameter λ according to Scheutjens-Fleer theory [39]. *r* is the chain length.

The value for A_2 has to be put into equation (1.43).

If two particles, each bearing an adsorbed layer of macromolecules, collide the adsorbed layers will tend to interpenetrate. The result is an increase in segment density in the overlapping volume together with conformational changes for the adsorbed macromolecules. Both effects contribute to the steric interaction (steric hindrance). They are often referred to as the "mixing" and "entropic" (or volume restriction) contribution, respectively [35—37].

During the approach of particles covered with macromolecules the adsorbed loops and tails lose configurational entropy. Hesselink derived approximate analytical expressions both for the entropic interaction assuming equal loops and for tails of different lengths, and also for the osmotic interaction again for equal sized loops and tails of different length.

The following set of equations was derived by Hesselink for neutral macromolecules:

$$\Delta\,^1V_{\text{steric}} = 2vk_{\mathrm{B}}T[Q(d)] + 2\left(\frac{2\pi}{9}\right)^{3/2} v^2 k_{\mathrm{B}}T(\alpha^2 - 1)\,\langle r^2\rangle\,[M(d)] \qquad (1.45)$$

$\Delta\,^1V_{\text{steric}}$ is the interaction energy per unit area, v the number of tails or loops, α the Flory (or expansion) parameter, and $\langle r^2\rangle^{1/2}$ is the root-mean-square end-to-end distance, defined as $\langle r^2\rangle^{1/2} = (il^2)^{1/2}$ (i number of segments with length l). The relationship between α and χ is shown by

$$\alpha^5 - \alpha^3 = \frac{27i^2 v_s^2 \left(\dfrac{1}{2} - \chi\right)}{(2\pi)^{3/2}\,v_1 (il^2)_\Theta^{3/2}} \qquad (1.46)$$

χ is the Flory-Huggins parameter ($\chi = 0.5$ in a so called "Θ-solvent"), v_s the segment volume, and v_1 the volume of a solvent molecule. The volume restriction effect, expressed by the function $Q(d)$, is given by the following equations:

equal loops: $$Q(d) \simeq 2(4\beta d^2 - 1)\exp(-2\beta d^2) \qquad (1.47)$$

equal tails: $$Q(d) \simeq 2\exp\left(-\frac{\beta d^2}{2}\right) \qquad (1.48)$$

tail size distribution: $$Q(d) \simeq 2\exp(-\beta d^2)^{1/2} \qquad (1.49)$$

The osmotic term is given by the following equations:

equal loops: $$M(d) \simeq \left(\frac{6\pi}{5}\right)^{1/2}\left(\frac{4}{5}\beta d^2 - 1\right)\exp\left(-\frac{2\beta d^2}{5}\right) \qquad (1.50)$$

equal tails: $$M(d) \simeq (3\pi)^{1/2}(2\beta d^2 - 1)\exp(-\beta d^2) \qquad (1.51)$$

tail size distribution:

$$M(d) \simeq 4[3\exp(-\beta d^2)^{1/2}]^{1/2}[(\beta d^2)^{1/2} - 2] + [\exp(-\beta d^2)^{1/2}][(\beta d^2)^{1/2} + 2] \qquad (1.52)$$

$$\beta = \frac{3}{\langle r^2\rangle} \qquad (1.53)$$

The main parameters determining the interaction energy of two approaching particles, each carrying an adsorbed polymer layer, are the average number of segments $\bar{i}$ per loop and the size distribution of tails, the number of loops or tails, the quality of the solvent, and the van der Waals interaction, in particular the contribution from the adsorbed layer.

The greater the solvation of the segments of the macromolecules, the lower will be the influence of the adsorbed layer on the total van der Waals interaction.

The interaction energy per unit area can be converted into the force between two spherical particles (radii a_1 and a_2) using Derjaguin's [38] expression

$$F = 2\pi\,^1V\sqrt{a_1 \cdot a_2} \qquad (1.54)$$

Below are summarized the main parameters determining the interaction between approaching particles each carrying an adsorbed neutral polymer layer:
— the average number of segments per loop,
— the size distribution of segments in the tails,
— the number of adsorbed loops or tails,
— the quality of the solvent,
— the contribution of the adsorbed layer to the van der Waals energy, which is generally small for the tails because of the low segment density in this layer (the higher the solvation of the segments the lower the contribution of the adsorbed layer to the van der Waals energy),
— the electrostatic forces between the particles which are in general screened because the decay in potential is largely within the adsorbed layer, particularly at higher ionic strength.

1.4.   Superposition of the Interparticle Forces

The stability of colloidal dispersions — or, alternatively, the rate of coagulation — depends on the interaction forces between approaching particles. Van der Waals attraction forces and electrostatic repulsion forces are nearly always present in every dispersion. The decay of the van der Waals forces with distance is proportional to $\frac{1}{d}$ and the decay of the electrostatic repulsion is proportional to $\ln \{1 + \exp [-\varkappa(r - 2a)]\}$ for the condition $\varkappa a > 5$. From this it follows that, at large distances, and at very short distances $V_A > V_{el}$. Therefore the total interaction energy—distance curve should in principle have two minima and one maximum. If direct contact occurred (i.e., the separation distance equals twice the appropriate atomic radii) then the primary minimum would be very deep. But this situation is unlikely to occur since the distance of closest approach is determined by adsorbed layers of counterions (in the Stern layer) and/or adsorbed solvent molecules. If we denote the diameter of the counterion or of the water molecules by x_δ then the distance of closest approach at particle contact is $2x_\delta$. The interaction energy at distance $2x_\delta$ (i.e., the primary minimum energy) determines therefore the properties of colloidal dispersions in respect to the reversibility or irreversibility of coagulation in the primary minimum.

Let us assume the distance of closest approach to be determined by twice the diameter of the counterions in the Stern layer. Then the electrostatic repulsion due to interpenetration of the diffuse double layers depends on the distance of closest approach $d = r - 2(a + x_\delta)$. This correction becomes important at small distances.

On the other hand adsorbed water molecules or counterions will not influence the van der Waals interaction, therefore V_A is proportional to $d = r - 2a$.

As was mentioned already in section 1.1., the electrostatic repulsion energy depends on whether the assumption is made of constant potential or constant surface charge during a collision. It is reasonable to suppose that the exchange of ions between the surface of the particles and the double layer is too slow to establish equilibrium within

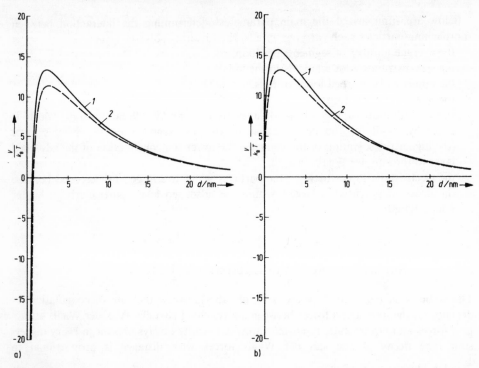

Figure 1.8
(a) Interaction energy-distance curve for spherical particles with $a = 10$ nm, $\psi_\delta = 50$ mV, $A = 10^{-19}$ J in 10^{-3} mol $\cdot$ dm^{-3} 1,1 electrolyte solution.
1, $\sigma_0 =$ const.; 2, $\psi_0 =$ const.
(b) The same as Figure 1.8a but with a distance of closest approach of 0.4 nm.

the time scale of the collision. Therefore it is preferable to calculate electrostatic repulsion assuming the condition of constant charge. But if we consider deaggregation then this process may take place after the equilibrium distributions of the ions had been established in the overlapping areas of the double layers. Therefore the constant potential approach is the better one. The differences in the total interaction distance curves for constant potential and constant surface charge are shown in Figure 1.8a for small particles of radii 10 nm and $\psi_\delta = 50$ mV and $A = 10^{-19}$ J, without introducing a distance of closest approach, and in Figure 1.8b on introducing a distance of closest approach of 0.4 nm. The calculations show that, in general, the repulsion energy is higher at constant charge, but the influence is not too large. However, the introduction of a distance of closest approach is very important in regard to the shape of the interaction energy-distance curve. As shown in Figures 1.8a and 1.8b a deep primary minimum is obtained only if the two particles come into direct contact. However, the introduction of a Stern layer which prevents the direct contact is enough to achieve stability. Therefore, we should expect in a real dispersion a state somewhere between these two limiting conditions.

As mentioned above, the boundary condition for rapid coagulation is given by

$$V_{\text{tot}} = V_A + V_{\text{el}} = 0 \tag{1.55}$$

Furthermore, if the maximum touches the x axis then the first derivative is also zero, i.e.,

$$\frac{\mathrm{d}V_{\text{tot}}}{\mathrm{d}d} = V'_{\text{tot}} = 0$$

or

$$V'_{\text{tot}} = V'_A + V'_{\text{el}} = 0 \tag{1.56}$$

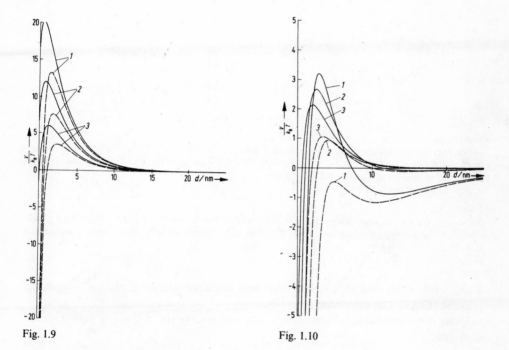

Fig. 1.9

Fig. 1.10

Figure 1.9
Interaction energy—distance curves for spherical particles with $a = 10$ nm, $A = 10^{-19}$ J in 10^{-2} mol $\cdot$ dm^{-3} 1,1 electrolyte solution and different Stern potentials.
Curves 1, $\psi_\delta = 60$ mV; curves 2, $\psi_\delta = 50$ mV; and curves 3, $\psi_\delta = 40$ mV with (solid lines) and without (dashed lines) a distance of closest approach of 0.4 nm.

Figure 1.10
Interaction energy-distance curves for spherical particles with $A = 10^{-19}$ J, $A_{\text{H}_2\text{O}} = 3.5 \cdot 10^{-21}$ J, $\psi_\delta = 30$ mV in 10^{-2} mol $\cdot$ dm^{-3} 1,1 electrolyte solution for different radii.
Curves 1, $a = 50$ nm; curves 2, $a = 25$ nm; curves 3, $a = 10$ nm with (solid lines), and without (dashed lines) a distance of closest approach of 0.4 nm.

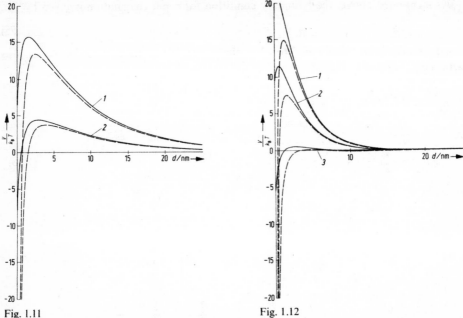

Fig. 1.11   Fig. 1.12

Figure 1.11
Interaction energy—distance curves for spherical particles with $A = 10^{-19}$ J, $a = 10$ nm, and different Stern potentials in 10^{-3} mol $\cdot$ dm^{-3} 1,1 electrolyte solution, with (solid lines) and without (dashed lines) a distance of closest approach of 0.4 nm,
1, $\psi = 30$ mV; 2, $\psi_\delta = 30$ mV.

Figure 1.12
Interaction energy—distance curves for spherical particles with $A = 10^{-20}$ J, $a = 50$ nm in 10^{-2} mol $\cdot$ dm^{-3} 1,1 electrolyte solution and with different Stern potential, solid lines with a distance of closest approach of 0.4 nm.
1, $\psi_\delta = 25$ mV; 2, $\psi_\delta = 20$ mV; 3, $\psi_\delta = 10$ mV.

Using both equations in combination with equation (1.7a) a critical coagulation concentration can be calculated.

Using the approximate equations (1.22) and (1.34) the following equations are obtained:

$$K \ln \left[1 + \exp \left(-\varkappa d \right) \right] = \frac{Aa}{12d} \tag{1.57}$$

and

$$K \frac{\varkappa \left[\exp \left(-\varkappa d \right) \right]}{1 + \exp \left(-\varkappa d \right)} = \frac{Aa}{12d^2} \tag{1.58}$$

this leads to the relationship

$$\varkappa_{\mathrm{cr}} \cdot d = \frac{\left\{ \ln \left[1 + \exp \left(-\varkappa_{\mathrm{cr}} d \right) \right] \right\} \left\{ 1 + \exp \left(-\varkappa_{\mathrm{cr}} d \right) \right\}}{\exp \left(-\varkappa_{\mathrm{cr}} d \right)} \tag{1.59}$$

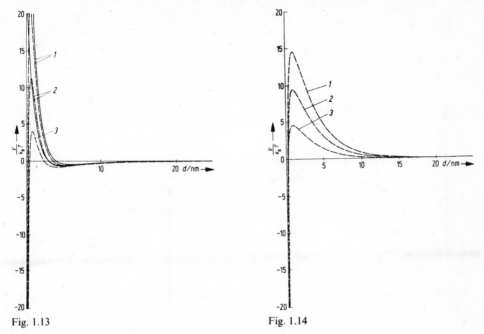

Fig. 1.13   Fig. 1.14

Figure 1.13
Interaction energy—distance curves for spherical particles with $A = 10^{-20}$ J, $a = 50$ nm in 10^{-1} mol · dm^{-3} 1,1 electrolyte solution and different Stern potentials (solid lines with a distance of closest approach of 0.4 nm).
$1, \psi_\delta = 35$ mV; $2, \psi_\delta = 30$ mV; $3, \psi_\delta = 25$ mV.

Figure 1.14
Interaction energy—distance curves for spherical particles with $A = 10^{-20}$ J, $\psi_\delta = 30$ mV and different radii in 10^{-2} mol · dm^{-3} solution.
$1, a = 30$ nm; $2, a = 20$ nm; $3, a = 10$ nm.

In order to obtain a general view of the interaction energy—distance curves for spherical particles of different types, some calculations are presented. The electrostatic repulsion was calculated, for small $\varkappa a$ using either equation (1.25) or (1.29), and for $\varkappa a > 5$ using equation (1.23). The van der Waals interaction was calculated with the equation derived by Clayfield, Lumb, and Mackey [27].

For metallic particles, such as gold, the Hamaker constant is about 10^{-19} J. For particles with radii of 10 nm in 10^{-2} mol · dm^{-3} 1,1 electrolyte solution, the energy—distance curves have been calculated for different Stern potentials. They are shown in Figure 1.9. Under the conditions chosen, no secondary minimum is obtained. A deep primary minimum is obtained if the distance of closest approach is not introduced. Slow coagulation should occur at Stern potentials lower than 50 mV. At the same electrolyte concentration and a Stern potential of 30 mV the influence of the particle radius is considered in Figure 1.10. The shape of the interaction energy distance curve depends strongly on the introduction of a distance of closest approach. For example, with particles of 50 nm radii rapid coagulation should occur in the primary minimum

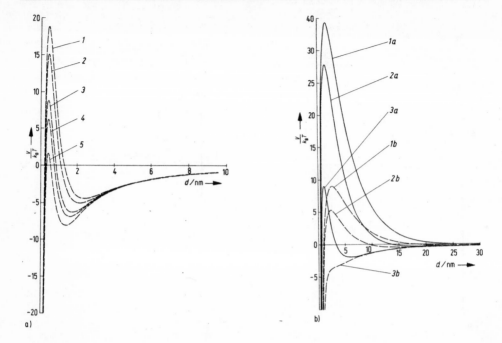

Figure 1.15
Interaction energy—distance curves for spherical particles with $A = 5 \cdot 10^{-21}$ J and $a = 200$ nm.

(a) $\psi_\delta = 25$ mV at different electrolyte concentrations.
1, $c = 0.3$ mol $\cdot$ dm^{-3}; 2, $c = 0.35$ mol $\cdot$ dm^{-3}; 3, $c = 0.45$ mol $\cdot$ dm^{-3}; 4, $c = 0.5$ mol $\cdot$ dm^{-3}; 5, $c = 0.6$ mol $\cdot$ dm^{-3}.

(b) $\psi_\delta = 15$ mV (solid lines) and 10 mV (dashed lines) at different electrolyte concentrations.
1a and 1b, $c = 5 \cdot 10^{-3}$ mol $\cdot$ dm^{-3}; 2a and 2b, $c = 10^{-2}$ mol $\cdot$ dm^{-3}; 3a and 3b, $c = 5 \cdot 10^{-2}$ mol $\cdot$ dm^{-3}.

if direct contact of the particles occurs. However, by introducing a distance of closest approach, coagulation should occur in the secondary minimum, followed by slow coagulation in the primary minimum. However, it is important to note the narrow range of the boundary conditions. The influence of the Stern potential on the interaction energy—distance curve for gold particles at lower electrolyte concentration is shown in Figure 1.11. It follows that even in 10^{-3} mol $\cdot$ dm^{-3} 1,1 electrolyte solution slow coagulation should occur in the primary minimum.

Similar calculations were performed with spherical particles having a Hamaker constant of about 10^{-20} J. Many oxide particles in water have a net Hamaker constant of this order. The influence of the Stern potential on the interaction energy—distance curves, at different electrolyte concentrations, is shown in Figures 1.12 and 1.13. As with gold particles, the primary minimum is supressed by introducing a distance of closest approach and coagulation should not occur. However, this is not in accord with experimental results. Therefore, subsequent calculations were performed without introducing a distance of closest approach. The influence of the size of the particles on the height of the energy barrier is shown in Figure 1.14.

In Figures 1.15a and 1.15b interaction energy—distance curves are shown for spherical particles with a Hamaker constant of $5 \cdot 10^{-21}$ J. This value is typical for latex particles. With particles of radii of 200 nm or greater a secondary minimum of sufficient depth for flocculation is formed. Therefore it is reasonable to infer that in many experiments described in the literature secondary minimum flocculation did occur. However, as shown in Figure 1.15, slow primary minimum coagulation is possible under certain conditions.

2.　　　Diffusion of Colloidal Particles

2.1.　　Translational Diffusion of Isolated Colloidal Particles

According to the kinetic theory of Maxwell and Boltzmann all molecules are in a state of continuous motion. Their translational kinetic energy only depends on their mass and on their velocity. In the law of equipartition of energy among the degrees of freedom all the molecules are taken to have the same translational kinetic energy, i.e.,

$$\frac{1}{2} m \bar{v}^2 = \frac{3}{2} k_B T \tag{2.1}$$

with m the mass of the molecules, $\bar{v}$ the average velocity, k_B the Boltzmann constant, and T the absolute temperature.

This equation also applies to the translational motion of colloidal particles. This motion of the particles was first observed by Brown, while using a microscope to investigate a dispersion of pollen grains in water. Hence this motion was later called Brownian motion.

It was shown by Graham that the diffusion of colloidal particles occurs at a much lower rate than that of molecules.

However, it is not possible to determine particle sizes from their direct observation in a microscope, because the displacement of the particles is the result of a "zig-zag" movement, which cannot be seen in the microscope.

The extremely irregular motion of colloidal particles is the result of collisions between particles and solvent molecules. The frequency of collisions in a dispersion with 10^{19} particles $\cdot$ cm^{-3} is about 10^{21} encounters per second.

The rotary diffusion of anisometric particles is also due to collisions of colloidal particles with solvent molecules.

In the absence of external fields and convection the mass m transferred across an area q during time dt as the result of a concentration gradient $\dfrac{dc}{dx}$ is described by Fick's first law:

$$\frac{1}{q} \frac{dm}{dt} = -D \frac{dc}{dx} = I \tag{2.2}$$

where q is the cross section, m is the mass, D is the diffusion coefficient, and I is the diffusion flux.

The variation of concentration with time at location x within the dispersion is described by Fick's second law:

$$\frac{dm}{dt} = -qD\left(\frac{\partial c}{\partial x} + \frac{\partial^2 c}{\partial x^2}\, dx\right) \tag{2.3}$$

When a particle moves within a liquid friction, occurs at the interface between the particle and the liquid. The friction becomes higher with increasing velocity of the particle. If the velocity of the particle increases, the viscous drag or resistance also increases. For a particle moving with velocity $\frac{dx}{dt}$ the frictional force is given by

$$F_R = f_s \frac{dx}{dt} \tag{2.4}$$

where f_s is the frictional coefficient.

The diffusion coefficient is related to the frictional coefficient through the following relationship:

$$D = \frac{k_B T}{f} \tag{2.5}$$

For spherical particles where the particle radius is much greater than the radius of the solvent molecules and under conditions of laminar flow, f_s is given by

$$f_s = 6\pi\eta a \tag{2.6}$$

with η the viscosity and a the particle radius.

The kinetic theory of Brownian motion was developed independently by Einstein [1], Von Smoluchowski [2], and Langevin [3].

The clearest derivation is that by Langevin; this is described in detail.

The starting point is the equation of linear motion of a particle with mass m, a driving force F, and a frictional force $\left(f_s \frac{dx}{dt}\right)$.

From Newton's law of motion the following relationship exists between the forces acting on the particle, its velocity $\frac{dx}{dt}$ and its acceleration $\frac{d^2x}{dt^2}$:

$$m\frac{d^2x}{dt^2} = F - f_s\frac{dx}{dt} \tag{2.7}$$

However, this expression cannot be used directly to find the mean displacement in the x direction. For mathematical reasons we have to consider the square of the velocity and the acceleration. This is because the probabilities that particles have velocity $+\left(\frac{dx}{dt}\right)$ and $-\left(\frac{dx}{dt}\right)$ and acceleration $+\left(\frac{d^2x}{dt^2}\right)$ and $-\left(\frac{d^2x}{dt^2}\right)$ are equal in both cases.

On differentiating x^2, the following expression for $\frac{dx}{dt}$ results:

$$\frac{dx}{dt} = \frac{1}{2x}\left(\frac{dx^2}{dt}\right) \tag{2.8}$$

Differentiating again leads to

$$\frac{d^2(x)^2}{dt} = \frac{d}{dt}\left(\frac{dx^2}{dt}\right) = 2x\left(\frac{d^2x}{dt^2}\right) + 2\left(\frac{dx}{dt}\right)^2 \qquad (2.9)$$

and

$$\frac{d^2(x)}{dt^2} = \frac{1}{2x}\left(\frac{d^2(x^2)}{dt^2}\right) + \left(\frac{dx}{dt}\right)^2 \qquad (2.10)$$

On substituting equations (2.8) and (2.10) into equation (2.7) we obtain

$$\frac{m}{2}\left(\frac{d^2(x)^2}{dt^2}\right) - m\left(\frac{dx}{dt}\right)^2 = Fx - \frac{f_s}{2}\left(\frac{d(x^2)}{dt}\right) \qquad (2.11)$$

Up to this point we have only made mathematical steps. The first physical consideration in averaging this expression for a large number of particles is implied in the assumption that the average kinetic energy for linear motion is expressed by the following equation:

$$m\left(\frac{dx}{dt}\right)^2 = k_B T \qquad (2.12)$$

The product of the driving force F and the displacement x is zero, on average, since there is an equal probability of collisions with solvent molecules occurring in any directions. Hence,

$$\overline{Fx} = 0 \qquad (2.13)$$

The equation for the motion of one particle is, therefore,

$$\frac{m}{2}\frac{d^2(x^2)}{dt^2} - k_B T = -\frac{f_s}{2}\frac{dx^2}{dt} \qquad (2.14)$$

Integration of this equation leads to

$$\frac{d(x^2)}{dt} = \frac{2k_B T}{f_s} - Ke^{-\frac{f_s t}{m}} \qquad (2.15)$$

K is an integration constant.

For sufficiently large observation times the second term in equation (2.15) is small compared to $\dfrac{2k_B T}{f_s}$. Hence, (2.15) reduces to

$$\frac{d(x^2)}{dt} = \frac{2k_B T}{f_s} \qquad (2.16)$$

Making use of equation (2.5) we obtain the following expression for the average square of the displacement:

$$\bar{x}^2 = 2Dt \qquad (2.17)$$

The time, $\dfrac{\dot{m}}{f_s}$, defined in equation (2.15) for spherical particles of radius $a = 10^{-5}$ cm and mass of 10^{-4} g in water at 293 K is of the order of 10^{-8} s. This time is small compared to observation time and, therefore, Einstein's equation (2.17) is valid for very long (virtually infinite) times.

2.2.   Translational Diffusion of Interacting Particles

2.2.1.   Translational Diffusion without Electrostatic Repulsion Forces

Von Smoluchowski considered the fast coagulation of colloidal particles to be determined only by the number of collisions of single particles with each other or with higher aggregates. He further assumed that collisions between particles or aggregates are only controlled by their Brownian diffusion. Von Smoluchowski defined a sphere of action with radius R where $R \simeq 2a$. Up to a separation distance R particles approach without any interaction. However, when this distance becomes smaller than R, the particles adhere irreversibly.

Von Smoluchowski's theory implied that there are three variables which influence the kinetics of coagulation: the particle radius a, the number of particles z_0, and the diffusion coefficient D. The dimensions of these variables are

$$D \sim \frac{l^2}{t}, \qquad z_0 \sim \frac{1}{l^3}, \qquad a \sim l$$

where l is the length and t is the time.

Because the dimension time only occurs in D and also because the extent of coagulation should be independent of time, the coagulation rate must depend on the product $D \cdot t$. Von Smoluchowski further showed that the influence of temperature on coagulation rate results directly from the temperature dependence of the diffusion coefficient [see equ. (2.5)]. Hence, increasing the viscosity reduces coagulation rates [see equ. (2.6.)]. When two (uncharged) particles approach within a certain distance the van der Waals attraction will become significant. One may define a "sphere of action" due to the van der Waals forces having radius R, such that all particle pairs at separations less than R will attract and coagulate. So the problem is to calculate the number of particle pairs for which this is the case. The area which encloses this sphere of attraction equals $4\pi R^2$ and therefore the number of single particles that pass through this sphere around one chosen particle in unit time is given by Fick's first law [equ. (2.2)],

$$\frac{\mathrm{d}v}{\mathrm{d}t} = 4\pi R^2 D \frac{\mathrm{d}z}{\mathrm{d}r} \tag{2.18}$$

where r is the center-to-center distance; v is the number of single particles passing through every sphere surrounding the central particle and finally hitting upon and fixing themselves to the central particle; and z is the particle concentration. $\dfrac{dz}{dr}$ has to be expressed as a function of the diffusion coefficient, the radius of the sphere of action, and time. This is achieved with Fick's second law [equ. 2.3)] as

$$\frac{dz}{dt} = D\left(\frac{\partial^2 z}{\partial x^2} + \frac{\partial^2 z}{\partial y^2} + \frac{\partial^2 z}{\partial z^2}\right) \tag{2.19}$$

In the case of spherical symmetry we obtain

$$\frac{dz}{dt} = D\left(\frac{\partial^2 z}{\partial r^2} + \frac{2}{r}\frac{\partial z}{\partial r}\right) \tag{2.20}$$

if the distance from the center of symmetry r is introduced instead of the coordinates x, y, z. The solution of this equation has the form

$$z = z_0 \left[1 - \frac{R}{r}\left(1 - \frac{2}{\sqrt{\pi}}\int_0^{\frac{r-R}{2\sqrt{Dt}}} \exp\left(-x^2\right) dx\right)\right] \tag{2.21}$$

In the stationary state, which is established almost at once, we can assume that the concentration at distance r is independent of time t, so that we obtain

$$z = z_0\left(1 - \frac{R}{r}\right) \tag{2.22}$$

The particle flux I towards the sphere of action of radius R (i.e., the number of particles that cross the surface of this sphere per second) is, according to Fick's first law [equ. (2.2)], given by

$$I = 4\pi R^2 D \left(\frac{\partial z}{\partial r}\right)_{r=R} \tag{2.23}$$

The derivative $\dfrac{\partial z}{\partial r}$ is calculated from equation (2.22)

$$\left(\frac{\partial z}{\partial r}\right)_{r=R} = \frac{z_0}{R} \tag{2.24}$$

and hence

$$I = 4\pi R D z_0 \tag{2.25}$$

This equation only holds for collisions with a particle that is fixed. This is obviously not the case, because this arbitrarily chosen particle has the same mobility as the particles that diffuse towards it. According to equation (2.17) the average square of the displacement equals

$$\bar{x}^2 = 2Dt$$

Hence the relative displacement between particle 1 and particle 2 is given by

$$D_{12} = \frac{\overline{(x_1 - x_2)^2}}{2t} = \frac{1}{2t}(\bar{x}_1^2 - 2\overline{x_1 x_2} + \bar{x}_2^2)$$

$$= \frac{\bar{x}_1^2}{2t} + \frac{\bar{x}_2^2}{2t} = D_1 + D_2 \qquad (2.26)$$

since the mean displacement $\overline{x_1 x_2}$ is equal to zero. For a monodisperse system we have

$$D_{11} = 2D_1 \qquad (2.27)$$

and, hence equation (2.25) for the flux of single particles towards a sphere of radius R that is also diffusing is now replaced by

$$I_{11} = 8\pi R D_1 z_0 \qquad (2.28)$$

The flux I_{11} of single particles equals the rate of aggregation with respect to one particle; clearly it will be z_0 times greater for all particles, i.e.,

$$\frac{dz_0}{dt} = -8\pi R D_1 z_0^2 = -2k_s z_0^2 \qquad (2.29)$$

where k_s represents the coagulation rate constant according to Von Smoluchowski. If we express D_1 by equations (2.5) and (2.6) and if we introduce $R = 2a$ then the following expression is obtained for k_s:

$$k_s = \frac{4}{3}\frac{k_B T}{\eta} \qquad (2.30)$$

Equation (2.29) may, in principle, be applied for all concentrations. It must be taken into consideration that in every encounter two primary particles will disappear. When both of the colliding particles are considered as central particles the collision frequency is doubled since each collision is counted twice.

2.2.2.   Translational Diffusion Including Hydrodynamic Interaction

In section 2.2.1. the relative diffusion of two particles was derived assuming that the frictional coefficient, f_s, [equ. (2.6)] remains independent of the separation distance and that the van der Waals forces, which are always present between approaching particles, are negligible. It was first mentioned by Derjaguin and Muller [6, 7] and later considered by Spielmann [8] and Honig et al. [9], that this assumption is not correct. It was shown that hydrodynamic interaction cannot be neglected.

The diffusional flux for the spherically symmetric, relative motion of two particles is given by the following equation:

$$I_{12} = I_D + I_F = 4\pi r^2 \left(D_{12}\frac{dz}{dr} + \frac{z}{f_s}\frac{dV}{dr}\right) \qquad (2.31)$$

I_{12} is the net radial flux, I_D is the sum of the relative diffusion flux, and I_F is the flux resulting from the interaction forces. Equation (2.31) holds also if the factor $2D_1$

is introduced, because the displacement and therefore the frictional force is also doubled.

The coefficient of relative diffusion, D_{12}, may be defined according to equation (2.27) only with the assumption that each of the particles diffuses independently in the absence of interaction forces. This is true only for distances large compared with their dimensions $(r \gg 2a)$. If particles approach to smaller distances then the frictional coefficient, defined by equation (2.6), is no longer constant, but changes with distance according to

$$f(r) = f_s \beta \left(\frac{r}{a} \right) \tag{2.32}$$

$\beta \left(\dfrac{r}{a} \right)$ is a dimensionless function with the boundary conditions

$$\beta = 1 \qquad \text{for } r \gg 2a \tag{2.32a}$$

$$\beta = \frac{1}{2} \frac{a}{r - 2a} \qquad \text{for } r - 2a \ll a \tag{2.32b}$$

A hyperbolic expression for the function β was given by Brenner [10] for the diffusion of spherical particles with equal radii.

It was shown by Honig and others [9] that the hydrodynamic correction can be approximated for large and small distances by the relation

$$\beta \simeq \frac{6u^2 + 13u + 2}{6u^2 + 4u} \tag{2.33}$$

where

$$u = \frac{r - 2a}{a} \tag{2.33a}$$

The solution of equation (2.31) should satisfy the following boundary conditions:

$$z = z_0 \quad \text{at} \quad r = \infty \tag{2.34}$$

and

$$z = 0 \quad \text{at} \quad r = 2a \tag{2.35}$$

With the equations (2.5), (2.22), and (2.24) together with the boundary condition $V(\infty) = 0$, we obtain equation (2.36),

$$z(r) = \exp\left[-\frac{V(r)}{k_B T} \right] \left[z_0 - \frac{I_{12}}{8\pi k_B T} \int_r^\infty \frac{f(r)}{r^2} \exp\left(\frac{V(r)}{k_B T} \right) dr \right] \tag{2.36}$$

Making use of the second boundary condition (2.35) we obtain the particle flux

$$I_{11} = \frac{8\pi k_B T z_0}{\displaystyle\int_{2a}^\infty \frac{f(r)}{r^2} \exp \frac{V(r)}{k_B T} dr} \tag{2.37}$$

If we introduce the relative diffusion coefficient through equations (2.5) and (2.6),

$$D_{12} = \frac{k_{B}T}{6\pi\eta a}$$

and the dimensionless relative distance u through equation (2.33a) and replace $f(r)$ by equation (2.32), we obtain

$$I_{11} = \frac{8\pi D_1 z_0}{\displaystyle\int_0^\infty \frac{\beta(u)}{(u + 2)^2} \exp\frac{V(u)}{k_{B}T}\ du} \tag{2.38}$$

$V(u)$ is the total interaction energy between approaching particles. From this equation it follows (making the hypothetical assumption that there are no attraction forces even at small separation distances) that

$$\beta(u) = \frac{1}{2u}$$

This implies that the particle flux would become zero, and no coagulation would occur, in the absence of electrostatic or steric repulsion forces. In this case the hydrodynamic resistance would prevent coagulation.

This is one of the most important conclusions for colloid chemists who want to prepare stable colloidal dispersions: they have to reduce the van der Waals attraction between interacting particles. This may be achieved by adsorption of low molecular or macromolecular surface active substances.

As was shown in chapter 1, the total interaction energy between particles is dependent on electrostatic repulsion, van der Waals attraction, and in some cases on steric repulsion. If we exclude steric repulsion and supress electrostatic repulsion by adding electrolytes until the energy barrier becomes zero ($V_{max} = 0$) we have the condition for rapid coagulation.

The diffusion flux then depends only on the van der Waals attraction and hydrodynamic interaction:

$$(I_{11})_{V_R=0} = \frac{8\pi D_1 z_0}{\displaystyle\int_0^\infty \frac{\beta(u)}{(u + 2)^2} \left[\exp\left(\frac{V_A(u)}{k_{B}T}\right) du\right]} \tag{2.39}$$

It was first mentioned by McGown and Parfitt [11] that in considering rapid coagulation in the primary minimum account has to be taken of the van der Waals attraction. The influence of the hydrodynamic interaction has been calculated by Honig and others [9]. They compared the diffusional flux according to the Smoluchowski equation (2.29) with the flux for rapid coagulation according to equation (2.39). They obtained the dimensionless rate constant

$$N_0 = \frac{(I_{11})_{V_R=0}}{I_{11}} = \frac{1}{\int\limits_0^\infty \frac{6u^2 + 13u + 2}{6u^2 + 4u} \cdot \frac{1}{(u+2)^2} \exp\left(\frac{V_A(u)}{k_B T}\right) du} \qquad (2.40)$$

$$N_0 \equiv \frac{2}{W}$$

The van der Waals energy was calculated using equation (1.32) for different values of the Hamaker constant. In Table 2.1 the influence of the hydrodynamic interaction is summarized. The dimensionless rate constant N_0 is compared with the rate constant (N_0') neglecting the hydrodynamic term [i.e., $\beta(u) = 1$, equ. (2.33)] for different Hamaker values. The results in Table 2.1 show that the influence of the hydro-

Table 2.1

Dimensionless rate constants for rapid coagulation, with and without the correction for hydrodynamic interaction (according to Honig, Roebersen, and Wiersema [9])

$A \cdot 10^{21}$ J	N_0	N_0'
0	0	2
0.25	0.768	2.018
0.5	0.821	2.029
1.0	0.882	2.043
2.5	0.976	2.082
5.0	1.061	2.128
10.0	1.160	2.193
25.0	1.320	2.313
50.0	1.467	2.433
100.0	1.644	2.590
250.0	1.922	2.842
500.0	2.173	3.076

dynamic interaction decreases with increasing Hamaker constant. The dimensionless rate constants are generally in the range of 40%—100% of the Smoluchowski values. In the absence of van der Waals attraction coagulation is not possible. The hydrodynamic interaction of two diffusing spherical particles of various sizes was considered in reference [23].

2.2.3.   Translational Diffusion in the Presence
              of Electrostatic Repulsion Forces — Weak Coagulation
              Due to an Energy Barrier

In this section the mutual diffusion of colloidal particles is considered when electrostatic repulsive forces are present. Under these conditions the interaction energy—
distance curves have a maximum (see Fig. 1.11), which prevents every encounter from

being effective. Only a fraction, $\dfrac{1}{W}$, of the collisions lead to aggregation [12].

With increasing repulsion the value of W increases. Therefore, W can be used as a quantitative measure of dispersion stability. The factor W is defined as the ratio of the
diffusion fluxes for rapid and slow coagulation, i.e., from equations (2.38) and
(2.39),

$$W \equiv \frac{(I_{11})_{V_R = 0}}{I_{11}} = \frac{\displaystyle\int_0^\infty \frac{\beta(u)}{(u+2)^2} \exp\left[\frac{V(u)}{k_B T}\right] du}{\displaystyle\int_0^\infty \frac{\beta(u)}{(u+2)^2} \exp\left[\frac{V_A(u)}{k_B T}\right] du} \tag{2.41}$$

According to this definition W cannot be less than 1. It should be mentioned, however,
that in many papers, following Fuchs [12], the diffusion flux for rapid coagulation
[equ. (2.39)] is considered without including van der Waals attraction or hydrodynamic
interactions. In that case,

$$W_{\text{Fuchs}} = 2 \int_0^\infty \exp\left[\frac{V(u)}{k_B T}\right] \frac{du}{(u+2)^2} = 2a \int_{2a}^\infty \exp\left[\frac{V(r)}{k_B T}\right] \frac{dr}{r^2} \tag{2.41a}$$

The neglect of V_A in equation (2.41 a) implies that W may now be less than 1.

  Since analytical calculations of W are not possible, Honig et al. [9] made a number
of numerical calculations. Some of their results are presented here. They calculated
the double layer repulsion using the following equation [30]:

$$V_R = G k_B T \exp(-\varkappa a u) \tag{2.42}$$

were G was introduced as

$$G = 8\varepsilon \left(\frac{k_B T}{ze}\right)^2 a \tanh^2\left(\frac{ze\psi_\delta}{4 k_B T}\right) \tag{2.42a}$$

The van der Waals interaction was calculated with equation (1.34), where the
parameter B is given by

$$B = \frac{A}{12 k_B T} \tag{2.43}$$

Table 2.2

Values of the parameter G [9]

ψ_δ in mV		G			
$z = 1$	$z = 2$	$a = 12$ nm	$a = 50$ nm	$a = 150$ nm	$a = 450$ nm
13.8	14.1	3	10	30	90
24.2	25.7	10	30	90	270
35.6	41.4	21	62	187	560
43.6	56.7	30	90	270	810
56.4	∞	47	140	420	1261
88.0	—	92	270	810	2430
∞	—	191	560	1681	5044

The relationship between ψ_δ and G for different electrolyte valences and different particle radii is shown in Table 2.2. It follows that for reasonable values of a and ψ_δ, G has values between ~ 10 and 810. The relation between G, $\varkappa a$, B, and log W for different Hamaker constants is shown in Table 2.3.

The dependence of log W on log $(\varkappa a)$ is shown in Figure 2.1a for different B values and $G = 90$.

The slope of the curve is approximately proportional to G and, therefore, to the particle radius, if ψ_δ and z are constant (Fig. 2.1 b).

This was shown experimentally to be the case by Reerink and Overbeek [13], for sols of silver iodide with different particle radii (Fig. 2.2). However, no influence of particle size on the slope was found for polystyrene latices by Ottewill and Shaw [14] (Fig. 2.3).

Table 2.3

Dependence of log W on the van der Waals attraction, electrostatic repulsion, and hydrodynamic correction [9]

G	$\varkappa a B$	log W		
		$B = 0.02$ $A = 10^{-21}$ J	$B = 0.2$ $A = 10^{-20}$ J	$B = 2$ $A = 10^{-19}$ J
10	2.51	0.22	0.52	—
	1.26	0.84	1.33	—
30	8.46	0.41	0.56	1.04
	0.95	7.60	7.85	8.46
90	28.46	0.91	1.09	1.57
	16.00	8.56	8.77	9.28
270	95.80	0.31	0.42	0.75
	76.10	9.81	10.01	10.46
810	287.4	2.90	3.09	3.52
	256.0	17.71	17.91	18.34

From the interaction energy—distance curves (Fig. 1.11) it can be seen that the maximum is located at separations of the order of a few tenths of a nanometer. Therefore the use of approximate equations for the interaction must be questioned.

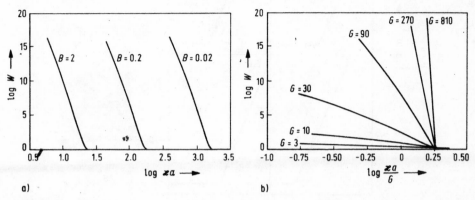

a) b)

Figure 2.1 a
Logarithm of delay factor as a function of log $\varkappa a$ at various values of B, G 90, hydrodynamic correction is included [9].

Figure 2.1 b
Logarithm of delay factor as a function of log $(\varkappa a/\sigma)$ at various values of the electrostatic repulsion.

For water at 298 K log $(\varkappa a/\sigma) = -0.53 \log z^3 - \log \tan h^2 \left(\dfrac{ze\psi_\delta}{4k_BT}\right) + \dfrac{1}{2} \log c$ [9].

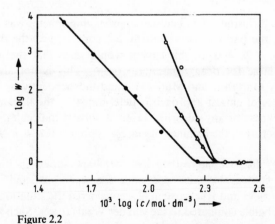

Figure 2.2
Stability ratio curves for silver iodide of different radii
$\bullet$, $a = 52$ nm; $\triangle$, $a = 200$ nm; $\bigcirc$, $a = 20{,}5$ nm coagulated with KNO_3 solutions. Electrolyte concentration in mol $\cdot$ dm^{-3} according to Reerink and Overbeek [13].

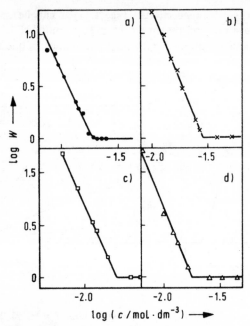

Figures 2.3
Stability ratio curves for monodisperse polystyrene latices of different size, (a) 30 nm, (b) 51.5 nm, (c) 121.2 nm, (d) 184 nm according to Ottewill and Shaw [14]; c, molar concentration of barium nitrate.

2.2.4.   Translational Diffusion under the Condition of Reversible Coagulation in the Primary or Secondary Minimum

In the previous section the diffusional flux between approaching spheres was calculated with the assumption that the particles come into direct contact, i.e., the distance of closest approach is $d = r - 2a = 0$. In this case it would seem that the interaction energy minimum is too large for deaggregation to occur. This is not always true, however, as was shown by Martynov and Muller [15] and independently by Frens and Overbeek [16]. The distance of closest approach is determined by the thickness of the Stern layer or by the thickness of any adsorbed layer of solvent molecules. Denoting the thickness of such a layer by δ, the distance of closest approach is $d = r - 2(a + \delta) = 2\delta$.

The interaction energy—distance curve then has the shape shown in Figure 1.11. The minimum energy is comparable with $k_B T$ or a few $k_B T$ units and, therefore, deaggregation will occur. Prieve and Ruckenstein [27] considered the situation when the double layer forces are negligible compared to the van der Waals forces. Such a situation arose in the limit of large ionic strength (rapid coagulation). Then W depends solely upon the ratio of $\dfrac{A}{k_B T}$ according to equation (2.41 a). It was shown that W equals unity only for $\dfrac{A}{k_B T} = 4.5$ (or for $A = 1.83 \cdot 10^{-20}$ J when $T = 300$ K). For smaller

values of $\dfrac{A}{k_B T}$ the stability ratio is greater than unity, indicating that the van der Waals forces are sufficiently weak that the viscous interaction between particles can lend a slight stability to the system. For $\dfrac{A}{k_B T} > 4.5$ the stability ratio becomes less than unity, indicating that migration of particles in the van der Waals force field is becoming important. As a result, the coagulation should become greater than that computed on the basis of diffusion alone and so W should become less than unity. However it will be shown later that this prediction was never confirmed experimentally. Another factor giving rise to a shallow primary energy minimum is a porous or a rough particle surface. In this case the amount of contact area between two nominately touching particles is reduced and hence so is the total van der Waals interaction energy (see section 1.2).

We now consider the particle flux for two approaching particles in the presence of a Stern layer.

The total particle flux depends on the flux of aggregation $(I_{12})_{V_R=0}$ as given by equation (2.37) with a change of the boundary condition:

$$(I_{12})_{V_R=0} = \frac{8\pi D_{12} z_0}{\displaystyle\int_{r=2(a+\delta)}^{r=\infty} r^{-2} \exp\left[\frac{V(r)}{k_B T}\right] dr} \tag{2.44}$$

This equation is also valid for the particle flux in the case of slow coagulation, i.e., in the presence of an energy barrier. Furthermore the hydrodynamic interaction can be introduced through equation (2.38). An analogous expression can be written for the deaggregation flux, I_{de}

$$I_{de} = \frac{8\pi D_{12} z_0}{\displaystyle\int_{r=2(a+\delta)}^{r=\infty} r^{-2} \exp\left[\frac{V^1(r)}{k_B T}\right] dr} \tag{2.45}$$

V^1 is the interaction energy at distance x from the primary minimum, that is

$$V^1 = V_x - V_{min}$$

The net particle flux $[I_{12} = (I_{12})_{V_R=0} - I_{de}]$ is then given by

$$I_{12} = 8\pi D_{12} z_0 \left\{ \frac{1}{\displaystyle\int_{2(a+\delta)}^{\infty} r^{-2} \exp\left[\frac{V(r)}{k_B T}\right] dr} - \frac{\exp\left(\dfrac{V_{min}}{k_B T}\right)}{\displaystyle\int_{2(a+\delta)}^{\infty} r^{-2} \exp\left[\frac{V(r)}{k_B T}\right] dr} \right\} \tag{2.46}$$

or

$$I_{12} = 8\pi D_{11} z_0 \left\{ \frac{1}{\int\limits_{2(a+\delta)}^{\infty} r^{-2} \exp\left[\frac{V(r)}{k_B T}\right] dr} \right\} \left[1 - \exp\left(\frac{V_{min}}{k_B T}\right) \right] \qquad (2.47)$$

The term in curly brackets in equation (2.47) describes rapid coagulation, according to equation (2.38), if hydrodynamic interactions are neglected.

In the absence of an energy barrier [i.e., $V(r)$ is determined only by the van der Waals attraction] the retardation factor, W, is related directly to the depth of the energy minimum, i.e.,

$$W = \left[1 - \exp\left(\frac{V_{min}}{k_B T}\right) \right]^{-1} \qquad (2.48)$$

Changing the boundary conditions in equation (2.47) according to Figure 1.15 one obtains the diffusional flux for secondary minimum coagulation [18, 19, 25—27]

$$|I_{12}|_{sec} = 8\pi D_{11} z_0 \left\{ \frac{1}{\int\limits_{r_s-2a}^{\infty} r^{-2} \exp\left[\frac{V(r)}{k_B T}\right] dr} \right\} \left[1 - \exp\left(\frac{V_{min}}{k_B T}\right)_{sec} \right] \qquad (2.49)$$

Bagchi modified equation (2.48) for secondary minimum:

$$W = \left\{ 1 - \exp\left[\frac{(V_{min} - k_B T)}{k_B T}\right] \right\}^{-1} \qquad (2.48\,a)$$

He assumed, arbitrarily, that if $|V_{min}| < k_B T$ the dispersion is thermodynamically stable and that only for $|V_{min}| > k_B T$ would any flocculation be observed.

2.2.5.   Translational Diffusion in Polydisperse Systems

The original theories of rapid and slow coagulation were developed under the assumption that all the particles have the same size. This is an oversimplification in two respects: firstly, most practical dispersions are polydisperse; secondly, even if they were monodisperse they would become polydisperse during coagulation. The theory of coagulation of polydisperse systems was first developed by Müller [20, 21].

The diffusional flux between two unequal, spherical particles for rapid coagulation is given by

$$I = 4\pi R_{12} D_{12} z_0$$

The sphere of interaction becomes

$$R_{12} = a_1 + a_2 \qquad (2.50)$$

(where a_1 and a_2 are particle radii) and the relative diffusion coefficient according to equation (2.27) is given by

$$D_{12} = D_1 + D_2$$

The diffusional flux for unequal spheres becomes

$$I_{12} = 4\pi z_0 (D_1 + D_2)(a_1 + a_2) \tag{2.51}$$

Expressing D_{12} by equations (2.5) and (2.6) leads to

$$I_{12} = \frac{2}{3} z_0 \frac{k_B T}{\eta} \left(\frac{1}{a_1} + \frac{1}{a_2} \right) (a_1 + a_2) \tag{2.52a}$$

or

$$I_{12} = \frac{2}{3} z_0 \frac{k_B T}{\eta} \frac{(a_1 + a_2)^2}{a_1 \cdot a_2} \tag{2.52b}$$

The Smoluchowski rate constant for unequal spheres is then given by

$$k'_S = \frac{1}{3} \frac{k_B T}{\eta} \frac{(a_1 + a_2)^2}{a_1 \cdot a_2} \tag{2.53}$$

For equal spheres $\dfrac{(a_1 + a_2)^2}{a_1 \cdot a_2}$ is 4, but for unequal spheres it is larger than 4. This means the collision probability increases with increasing polydispersity.

Recently [23] collision efficiencies of unequal spheres were computed taking into account hydrodynamic, van der Waals, and double layer interactions between two approaching particles.

2.2.6.   Translational Diffusion of Nonspherical Particles

The diffusional flux of particles having any shape is given by the general equation (2.25). The problem consists in determining expressions for R_{12} and D_{12}. When two general-shaped particles approach, interaction will start at different separations, depending on their mutual orientation. For example two rodlike particles may approach with their long axes inclined at a fixed angle to the normal of the collision nucleus (Fig. 2.4). In this case the separation of the two centers of mass is fixed at any position. Therefore one can ask the question: how many particles will coagulate in that special orientation? Coagulation will occur if the centers of mass reach a distance r where the van der Waals attraction becomes dominant. However, these distances are different at different points of the interface of the central particle. All these points are placed at the interface σ around the central particle (Fig. 2.4).

The diffusional flux through this surface element is defined by the surface integral

$$I_{12} = \int_\sigma D_{12} \, \text{grad} \, z \, d\sigma \tag{2.54}$$

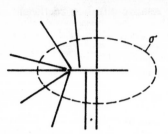

Figure 2.4
Scheme of coagulation of cylindrical particles
according to Müller [20].

Every particle entering this area becomes coagulated. In other words, $z = 0$ within this volume.

Following Müller [22], we introduce equivalent radii a_1^*, a_2^* for anisometric particles corresponding in each case to a sphere of equal mobility as the anisotropic particle.

From equation (2.52) we obtain for the diffusional flux

$$I_{12} = \frac{2}{3} z_0 \frac{k_B T}{\eta} \frac{(a_1^* + a_2^*)^2}{a_1^* \cdot a_2^*} \cdot C \tag{2.55}$$

where C is called the "shape factor".

If we compare the diffusional flux of different shaped particles with the flux of equal spheres I_{11} [equ. (2.29)], we obtain

$$I_{12} = I_{11} \cdot \frac{1}{4} \frac{(a_1^* + a_2^*)^2}{a_1^* \cdot a_2^*} \cdot C \tag{2.56}$$

and for the change of the total number of particles with time

$$\frac{dz_0}{dt} = \frac{2}{3} z_0^2 \frac{k_B T}{\eta} \frac{(a_1^* + a_2^*)^2}{a_1^* \cdot a_2^*} \cdot C \tag{2.57}$$

The shape constant C was analyzed by Müller for two kinds of particles: cylinders and platelets. Cylinders have one long axis of rotation b and two small axes c. Rotation around axis b is more likely than around c owing to the lower hydrodynamic resistance.

Let us assume for simplicity that the axis of one particle is normal to the surface of the central particle. The shape factor for cylinders, for the condition $\frac{b}{c} \gg 1$, is

$$C = 1.49b$$

and the equivalent radii equal

$$a^* = \frac{2}{3} \frac{b}{\ln 2 \left(\frac{b}{c}\right)}$$

along the axes and

$$a^* = \frac{4}{3} \frac{b}{\ln 2 \left(\dfrac{b}{c}\right)}$$

across the rotational axes. Platelet particles have one small axis b and two long cross axes c. The shape factor, for the condition $\dfrac{b}{c} \ll 1$, is

$$C = \frac{3\sqrt{3}}{c}$$

and the equivalent radii are

$$a^* = \frac{16}{9\pi} c$$

in the direction of cross axis and

$$a^* = \frac{16}{3\pi} c$$

along the cross axis.

O'Neill et al. [28] have shown that for doublets of spheres the diffusion coefficient is orientationally dependent and has two components, the diffusion coefficient in the line joining the centers of the spheres

$$D_1 = \frac{k_B T}{6\pi\eta a \; 1.290}$$

and perpendicular to this direction

$$D_2 = \frac{k_B T}{6\pi\eta a \; 1.449}$$

The equivalent radii for cylinders and for platelets are larger for the movement across the rotational axis than for those along the axis.

The diffusional flux of two particles having any shape can be approximated by introducing the cube root of their volumes instead of their radii. One then obtains, instead of equation (2.55),

$$I_{12} = \frac{2}{3} z_0 \frac{k_B T}{\eta} \frac{(v_1^{1/3} + v_2^{1/3})^2}{v_1^{1/3} \cdot v_2^{1/3}} \tag{2.58}$$

2.2.7. Direct Measurements of Translational Diffusion Coefficients

An apparatus used for the direct microscopic observation of particles in Brownian motion has been described in ref. [24]. The apparatus consisted of a light source which was collimated and passed through a heat filter prior to illuminating the observation cell. This cell was set in a framework which allowed for simultaneous movement of the cell in three mutually perpendicular directions. The movement of an aggregate was tracked by moving the observation cell and keeping the recording device stationary. The observation cell was viewed through a microscope which had a range of available magnifications from $40 \times$ to $640 \times$. The eyepiece lens was fitted with a squared graticule in order that a positive location point be available.

The motions of the particles or aggregates, respectively, were recorded with a high-speed intermittent pin registered cinematographic camera. The camera was equipped with a periscope viewing device enabling simultaneous viewing and recording of particle movement.

The observation cell was immersed in a water jacket. The particle number concentration of approximately 10^7 cm^{-3} was used. After developing, the film was digitized and analyzed. The experiments were performed in a mixture of D_2O:water (1.083), but the diffusion coefficients were corrected and represent the diffusion coefficients in pure water at 298 K.

The results for aggregates of polystyrene particles of 1 μm radius are tabulated (Table 2.4).

Table 2.4

Experimentally determined translational diffusion coefficients of polystyrene particles ($a = 1$ μm) and their aggregates in water at 298 K

z_0 of primary particles	Type of aggregate	Experimental translational diffusion coefficient	
		in 10^{-9} cm$^2 \cdot$ s^{-1}	Error in %
1	—	2.43	1.0
2	—	1.76	1.5
3	linear	1.56	1.5
3	cluster	1.47	3.0
4	linear	1.34	5.0
4	cluster	1.09	9.0
5	linear	0.98	7.5
5	cluster	0.80	18.0
6	linear	0.87	7.0
6	cluster	0.76	17.0

The values of the translational diffusion coefficients for aggregates show that there is a rapid decrease in the mobility of aggregates with increasing aggregate size. The results show further that the linear aggregates have higher diffusion coefficients and are therefore more mobile than the compact aggregates. It would be expected that the greater the degree of sphericity of an aggregate the smaller the drag coefficient and thus the greater the mobility. Hence the experimental results are contrary to the expected results. These results reflect that linear aggregates show a higher mobility in the direction of the line of centers.

The difficulty in describing the diffusion coefficient of aggregates lies in the complexity of the hydrodynamic resistance for the fluid flow in and around an aggregate.

3. Coagulation Kinetics

3.1. Von Smoluchowski's Theory of Rapid Coagulation

The number of collisions of singlet particles with other singlet particles follows from equation (2.29):

$$\frac{dz_1}{dt} = -4\pi R_{11} D_{11} z_1^2 \tag{3.1}$$

This equation corresponds to a bimolecular chemical reaction[1]). The concentration of single particles at the time t is then obtained using

$$z_1 = \frac{z_0}{1 + 2\pi R_{11} D_{11} z_0 t} = \frac{z_0}{1 + 4\pi R_{11} D_1 z_0 t} = \frac{z_0}{1 + k_s z_0 t} \tag{3.2}$$

$k_s = 2\pi R_{11} D_{11}$ is referred to as the Smoluchowski rate constant [equ. (2.30)].

However, changes in the concentration of singlet particles are not only from collisions with other singlets but also from collisions with aggregates (doublets, triplets, etc.). Therefore the total change in singlet concentration is given by

$$\frac{1}{4\pi} \frac{dz_1}{dt} = -R_{11} D_{11} z_1^2 - R_{12} D_{12} z_1 z_2 - R_{13} D_{13} z_1 z_3 \cdots - R_{1i} D_{1i} z_1 z_i$$

$$= -z_1 \sum_{i=1}^{\infty} R_{1i} D_{1i} z_i \tag{3.3}$$

The change of the number of doublets is given by

$$\frac{1}{4\pi} \frac{dz_2}{dt} = \frac{1}{2} R_{11} D_{11} z_1^2 - R_{12} D_{12} z_1 z_2 - R_{22} D_{22} z_2^2 - R_{2i} D_{2i} z_2 z_i$$

$$= \frac{1}{2} R_{11} D_{11} z_1^2 - z_2 \sum_{i=1}^{\infty} R_{2i} D_{2i} z_i \tag{3.4}$$

and for triplets

$$\frac{1}{4\pi} \frac{dz_3}{dt} = R_{12} D_{12} z_1 z_2 - R_{13} D_{13} z_1 z_3 - R_{23} D_{23} z_2 z_3 - R_{33} D_{33} z_3^2 - R_{3i} D_{3i} z_3 z_i$$

$$= R_{12} D_{12} z_1 z_2 - z_3 \sum_{i=1}^{\infty} R_{3i} D_{3i} z_i$$

---

[1]) In every collision two single particles will disappear. But it is not necessary to multiply the right hand side of equation (3.1) by a factor of 2, because every particle is counted once as the central particle (see section 2.2.1.).

In general, the change of the number of aggregates of kind k is given by

$$\frac{1}{4\pi} \frac{dz_k}{dt} = \frac{1}{2} \sum_{\substack{i=1 \\ j=k-i}}^{j=k-1} R_{ij} D_{ij} z_i z_j - z_k \sum_{.i=1}^{\infty} R_{ik} D_{ik} z_i \tag{3.5}$$

Summing over all aggregate types, we obtain

$$\frac{1}{4\pi} \frac{d\sum_{1}^{\infty} z_i}{dt} = \frac{1}{2} \sum_{i=1}^{\infty} R_{ii} D_{ii} z_i^2 + \sum_{\substack{i=1 \\ j=i+1}}^{\infty} R_{ij} D_{ij} z_i z_j - \sum_{i=1}^{\infty} \sum_{j=1}^{\infty} R_{ij} D_{ij} z_i z_j$$

$$= -\frac{1}{2} R_{ii} D_{ii} \left(\sum_{i=1}^{\infty} z_i \right)^2 \tag{3.6}$$

The following, simplifying assumptions were made by von Smoluchowski:

1. All aggregates were approximated by spheres. This is strictly correct for coalescing particles. Making this assumption, however, we obtain,

$$D_{ij} = D_i + D_j \quad \text{and} \quad R_{ij} = R_i + R_j$$

and from equation (2.53) the Smoluchowski rate constant for unequal spheres is given by

$$k_s = \frac{1}{3} \frac{k_B T}{\eta} \frac{(a_1 + a_2)^2}{a_1 \cdot a_2}$$

This value is larger than the Smoluchowski rate constant for equal spheres

$$\frac{(a_1 + a_2)^2}{a_1 \cdot a_2} > 4 \quad \text{if } a_1 \neq a_2$$

2. The products of the relative diffusion coefficients, D_{ij}, and the spheres of interaction, R_{ij}, are equal, that is

$$D_{ij} \cdot R_{ij} = 2D_1 R$$

D_1 is defined by equation (2.28).
This assumption is correct only at the early stages of coagulation, when equation (3.6) may be integrated. We obtain for the change of the total number of particles

$$\sum z = z_1 + z_2 + z_3 + \ldots = \frac{z_0}{1 + k_s z_0 t} \tag{3.7}$$

For the number of single particles, doublets, and triplets the following equations are obtained

$$z_1 = \frac{z_0}{(1 + k_s z_0 t)^2}$$ (3.8)

$$z_2 = \frac{z_0(k_s z_0 t)}{(1 + k_s z_0 t)^3}$$

$$z_i = \frac{z_0(k_s z_0 t)^{i-1}}{(1 + k_s z_0 t)^{i+1}}$$

The coagulation time, T_{ag}, is given by

$$T_{ag} \equiv \frac{1}{4\pi D_1 R_{11} z_0} = \frac{1}{k_s z_0}$$ (3.9)

At time T_{ag} the total number of particles equals $1/2\, z_0$, the number of single particles $1/4\, z_0$, and the number of dimer particles is given by

$$z_i = \frac{z_0}{2^{(i+1)}}$$ (3.10)

In Figure 3.1 the decreases in the total number of particles, single particles, doublets, triplets, and quadruplets with time are shown. It is seen that the curves for the

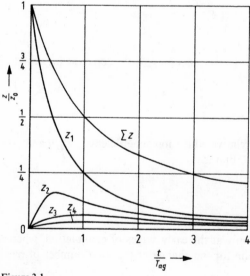

Figure 3.1
The relative change of particle concentration with time according to Von Smoluchowski theory.

number of doublets, triplets, and quadruplets reach a maximum at time

$$t_{max} = \frac{i-1}{2} T_{ag}$$ (3.11)

3.2.   Corrections to the Von Smoluchowski Theory for Polydispersity

As was shown in section 2.2.4. and in the previous section, polydispersity enhances coagulation. The influence of polydispersity has been investigated by many authors both experimentally and theoretically [3—9]. It was assumed by Lee [9] that the initial size distribution and the time dependent size distribution during coagulation can be represented by a log-normal function.

The complete size distribution of a coagulating colloidal dispersion can be described by the following nonlinear integro-differential equation, if the particle size distribution is taken to be continuous [8, 9]:

$$\frac{\partial z(vt)}{\partial t} = \frac{1}{2} k_s \left[\int_0^v \beta(v_1 - v_2, v_2) \, z(v_1 - v_2, t) \, z(v_2, t) \right.$$

$$\left. - z(v_1, t) \int_0^\infty \beta(v_1, v_2) \, z(v_2, t) \, dv_2 \right]$$ (3.12)

$\beta(v_1, v_2)$ is a normalized collision kernel, defined by equation (2.58)

$$\beta(v_1, v_2) = \frac{(v_1^{1/3} + v_2^{1/3})^2}{v_1^{1/3} \cdot v_2^{1/3}}$$ (3.13)

There exists no analytic solution of equation (3.12). The size distribution function for particles with volume v and a log-normal distribution is written as

$$z(v, t) = \frac{z}{3 \sqrt{2\pi} \ln \sigma} \exp \left[\frac{-\ln^2 \left(\frac{v}{v_g} \right)}{18 \ln^2 \sigma} \right] \frac{1}{v}$$ (3.14)

$v_g(t)$ is the number median particle volume and $\sigma(t)$ is the geometric standard deviation based on particle radius. The three parameters of the distribution z, σ, and v_g are allowed to vary with time. A basic assumption is the correlation of the true distribution with a log-normal distribution; there is some support from the literature [10—12] that this assumption is justified. Lee has derived the following equations for the relation between v_g and σ:

$$v_g = v_g^0 \left[\frac{\exp(9 \ln^2 \sigma_0) - 2}{\exp(9 \ln^2 \sigma) - 2} \right] \exp \left[\frac{9}{2} (\ln^2 \sigma_0 - \ln^2 \sigma) \right]$$ (3.15)

v_{g0} and σ_0 are the initial values of v_g and σ, respectively. For the relation between the number of particles and the standard deviation, equation (3.16) was obtained

$$\ln^2 \sigma = \frac{1}{9} \ln \left[2 + \frac{\exp (9 \ln^2 \sigma_0) - 2}{1 + [1 + \exp (\ln^2 \sigma_0)] k_s z_0 t} \right] \tag{3.16}$$

If equation (3.16) is substituted into equation (3.15), we obtain

$$\frac{v_g}{v_g^0} = \frac{\exp \left(\frac{9}{2} \ln^2 \sigma_0 \right) \{1 + [1 + \exp (\ln^2 \sigma_0)]\} k_s z_0 t}{\left[\dfrac{2 + \{\exp (9 \ln^2 \sigma_0) - 2\}}{\{1 + [1 + \exp (\ln^2 \sigma_0)] k_s z_0 t\}} \right]^{1/2}} \tag{3.17}$$

Finally the following equation for the decrease of the total number of particles is obtained

$$\frac{z}{z_0} = \frac{1}{1 + [1 + \exp (\ln^2 \sigma_0)] k_s z_0 t} \tag{3.18}$$

The quantity in brackets in the denominator corrects the collision constant for polydispersity. Equations (3.18) and (3.7) become identical when $\sigma_0 = 1$.

The decay of the particle number z, divided by the initial number of particles, is shown in Figure 3.2 as a function of the reduced time, for various geometric standard deviations, σ_0. These calculations have shown that polydispersity has a pronounced effect on coagulation kinetics. For an initially monodisperse sample the geometric standard deviation will increase from 1.0 to 1.32. If the change in particle number is calculated according to Smoluchowski's equation (3.7) or Lee's equation (3.18) the difference between these two values is less than 4%. With an initially polydisperse system the differences between both equations become larger, the greater the difference in the geometric standard deviations. Lee further showed that, independently of the initial standard deviation (Fig. 3.2), the size distribution approaches that for a geometric standard deviation of σ_∞.

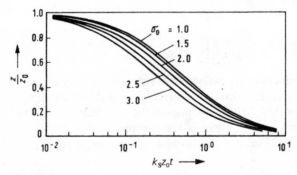

Figure 3.2
Decrease in the total number of particles z/z_0 as a function of dimensionless time for several standard deviations σ_0 according to Lee [9].

This result can be compared with the self-preserving distribution predicted by Friedlander and Wang [6]. They showed that the decay of the total number of particles is given by

$$\frac{z}{z_0} = \frac{1}{1 + (1 + 1.1289)\, k_s z_0 t} \tag{3.19}$$

Lee obtained the equation

$$\frac{z}{z_0} = \frac{1}{1 + (1 + 1.080)\, k_s z_0 t} \tag{3.20}$$

The difference in z predicted from these equations is about 1.5%.

These calculations have shown that von Smoluchowski's approximate theory may be used with reasonable accuracy for initially monodisperse systems, and also for dispersions having low polydispersity initially.

3.3. Correction to the Von Smoluchowski Theory for Slow Coagulation

It was assumed in the previous section that every collision is effective. However, this may not be the case for several reasons. For example, if the energy barrier between approaching particles is of the order of a few $k_B T$, then not every collision will be effective. In this case a factor α_{ef} may be introduced to account for the efficiency of collisions (collision probability). In the same way a deaggregation probability factor β may be introduced to account for reversible coagulation in a shallow primary or secondary minimum. Both factors can be expressed in terms of the interaction energy between the particles. If we consider the formation of doublets during rapid coagulation, then the decrease of single particles with time is given by equation (3.1):

$$-\frac{dz_1}{dt} = 4\pi R_{11} D_{11} z_1^2 \equiv \alpha_{1,\,col}\, z_1^2 \tag{3.21}$$

If we now introduce the aggregation probability defined by

$$\alpha_{1,\,ef} = \frac{1}{W} \tag{3.22}$$

than equation (3.21) becomes

$$-\frac{dz_1}{dt} = \alpha_{1,\,col}\, \alpha_{1,\,ef}\, z_1^2 \tag{3.21a}$$

and with W defined by equation (2.41 a), equation (3.21) becomes

$$-\frac{dz_1}{dt} = \frac{4\pi D_{11} z_1^2 (a + \delta)}{\int\limits_0^\infty \frac{du}{(u + 2)^2} \exp\left(\frac{V(u)}{k_B T}\right)}$$

$$\equiv \frac{4\pi D_{11} z_1^2}{\int\limits_{2(a+\delta)}^\infty \frac{dr}{r^2} \exp\left(\frac{V(r)}{k_B T}\right)} \tag{3.23}$$

note that the factor $2(a + \delta)$ represents the sphere of interaction, and δ is the thickness of the Stern layer. In reality the sphere of interaction, R, is usually somewhat larger than $2(a + \delta)$, depending on the particle radius and on the Hamaker constant. If we also introduce the hydrodynamic interaction then equation (3.23) takes the form,

$$-\frac{dz_1}{dt} = \frac{4\pi D_{11} z_1^2}{\int\limits_{2(a+\delta)}^\infty \frac{\beta(r)}{r^2} dr \left(\exp \frac{V(r)}{k_B T}\right)} \tag{3.24}$$

3.4.   Introduction of Reversibility into Coagulation Kinetics

As was described in sections 1.4. and 2.2.4., coagulation in the primary minimum is not necessarily irreversible. Retardation of coagulation kinetics may be caused by de-aggregation. This effect was first considered by Martynov and Muller [13]. They calculated the change with time of the number of particles of kind j through aggregation, i.e.,

$$\left(\frac{dz_j}{dt}\right)_{ag} = \frac{1}{2} \sum_{i=1}^\infty \alpha_{i,\,j-1,\,col} \alpha_{i,\,j-1,\,ef} z_i z_j - z_j \sum_{i=1}^\infty \alpha_{ij,\,col} \alpha_{ij,\,ef} z_i \tag{3.25}$$

The change with time of the number of particles of kind j by deaggregation is dependent on the deaggregation of particles of kind j and on deaggregation of higher aggregates, which leads to the formation of particles of kind j, i.e.,

$$\left(\frac{dz_j}{dt}\right)_{de} = -\beta_j z_j + \sum_{k=1}^\infty \beta_{jk} z_{jk} \tag{3.26}$$

where $\beta_j = \sum\limits_{i=1}^{j-1} (1 + d_{i,\,j-i}) \beta_{i,\,j-i}$ and $\beta_{i,\,j-1} = \beta_{j-i,\,i}$

($d_{i,\,j-i}$ is the Kronecker symbol.) We can now define the aggregation time $T_{ij,\,ag}$ and deaggregation time $T_{ij,\,de}$ in accord with equation (3.9):

$$T_{ij,\,ag} = \frac{1}{2\pi D_{ij} R_{ij} \alpha_{ij,\,ef} z_0} = \frac{W_{ij}}{2\pi D_{ij} R_{ij} z_0} \tag{3.27}$$

Under equilibrium conditions, the rate of aggregation equals the rate of deaggregation,

$$I_{ag} = I_{de} = \frac{4\pi R_{ij} D_{ij} z_j}{W_{ij}}$$

and the degree of deaggregation is therefore

$$\beta = \frac{I_{de}}{N_{ij,ag}} \tag{3.28}$$

and

$$N_{ij} = n_j \int_{r-2a}^{\infty} \left[\exp\left(\frac{V_{ij}(r)}{k_B T}\right) - 1 \right] 4\pi r^2 \, dr = 4\pi R_{ij}^3 n_j \Gamma_{ij} \tag{3.29}$$

where $4\pi R_{ij}^3$ is the interaction volume

$$\Gamma_{ij} = \int_{x_{ij}}^{\infty} \left[\exp\left(E_{ij}(x)\right) - 1 \right] x^2 \, dx \tag{3.30}$$

and

$$x_{ij} = \frac{d_{ij}}{R_{ij}} \simeq 1 \quad \text{and} \quad \beta_{ij} = \frac{D_{ij}}{R_{ij} W_{ij} \Gamma_{ij}} \tag{3.31}$$

where

$$\frac{R_{ij}^2}{D_{ij}} \equiv T_{ij,de}$$

$T_{ij,de}$ is called the deaggregation time. From this equation it follows that the higher the barrier and the deeper the minimum, the smaller the deaggregation coefficient.

If we now consider the formation of doublets from singlet particles and their breakup then

$$\frac{dz_1}{dt} = -\alpha_{col}\alpha_{ef} z_1^2 + 2\beta z_2 \tag{3.32}$$

and

$$\frac{dz_2}{dt} = \frac{1}{2}\alpha_{col}\alpha_{ef} z_1^2 - \beta z_2 \tag{3.33}$$

at equilibrium

$$\frac{d\Sigma z}{dt} = \frac{dz_1}{dt} + \frac{dz_2}{dt} = 0 \tag{3.34}$$

$$-\frac{1}{2}\alpha_{col}\alpha_{ef} z_1^2 + \beta z_2 = 0 \tag{3.35}$$

If we replace α_{col} by equation (3.21), α_{ef} by equations (2.41) and (3.22), and β by (3.31), we obtain

$$-\frac{1}{2}\frac{4\pi R_{12}D_{12}z_1^2}{W_{12}} + \frac{D_{12}z_2}{R_{12}^2 W_{12}\Gamma_{12}} = 0 \tag{3.36}$$

$$\frac{\alpha}{\beta} = \frac{2z_2}{z_1^2} = 4\pi R_{12}^3\Gamma_{12} = 4\pi R_{12}^3 \int\limits_{x_{12}}^{\infty} [\exp(E_{12}(x)) - 1]\, x^2\, dx \tag{3.37}$$

In the lower limit, $x_{12} = 2a$ (twice the particle radius).

The interaction energy E_{ij} here is defined in the context of the breakup of doublets. For higher aggregates the mode of breakup depends on the structure of the aggregate. For example, let us consider the deaggregation of a single particle from a quadruplet as shown schematically in Figure 3.3. In schemes (a) and (b) only one contact is broken (linear aggregate), in scheme (c) two, and in scheme (d) three contacts are broken if particle 4 deaggregates. However, the energy minimum is deeper in this orientation (compact aggregates). With increasing aggregate size a decrease of the breakup probability is to be expected.

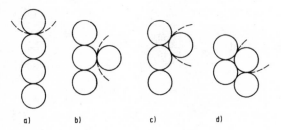

a)   b)   c)   d)

Figure 3.3
Scheme of coagulation of spherical particles in linear and compact aggregates.

Now let us formulate the criterion for rapid coagulation. The coagulation rate is a maximum if

$$T_{ij,\,ag} \ll T_{ij,\,de} \tag{3.38}$$

i.e., $V_{ij} < 0$ at all separations.

The rate increases for higher aggregates since

$$|V_{ij}| \gtrless |V_{11}| \tag{3.39}$$

However, this inquality does not necessarily correspond to the maximum coagulation rate since if

$$T_{ij,\,de} < T_{ij,\,ag} \tag{3.40}$$

then deaggregation can offset (in part) any aggregation, and the dispersion may become stable. Therefore, the second criterion for rapid coagulation is

$$T_{ij,\,de} \gg T_{ij,\,ag} \tag{3.41}$$

or, using equation (3.29),

$$4\pi R_{ij}^3 n_j \Gamma_{ij} \gg 1 \tag{3.42}$$

The derivation of equation (3.29) did not include the fact that the total interaction energy is the superposition of electrostatic and van der Waals energy. It is of general applicability and is valid, even if structural (steric) or magnetic forces (as in ferrofluids) are present.

Consider the case of linear aggregation (Figs. 3.3a, 3.3b), i.e., where every particle has contact with only one other particle. An alternative to linear aggregates are compact aggregates in which each particle has more than one contact (Figs. 3.3c and 3.3d). In this case the total interaction energy is about 2—3 times higher than in a linear aggregate. Therefore one might expect that compact aggregates are formed preferentially.

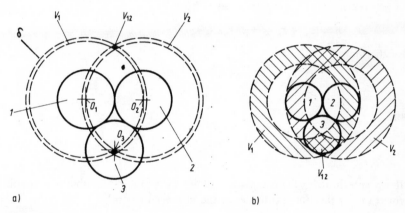

Figure 3.4
Scheme of coagulation of spherical particles. (a) separation distance δ much smaller than a (latexes); (b) $\varkappa a \approx 1$ (metallic particles).

However, other factors have to be considered [14]: If the separation distance δ, beyond which the interaction forces between any two particles are negligible, is small in comparison to the particle radius, i.e.,

$$\delta \ll a \tag{3.43}$$

the the volumes v_1 and v_2 (see Fig. 3.4) in which the center O_3 of the particle 3 can be located if this particle interacts either with particle 1 or particle 2 are much greater than the interaction volume v_{12} in which the center of particle 3 can be located if this particle interacts with both particles, e.g.,

$$v_{12} \ll v_1 + v_2 \tag{3.44}$$

For this reason the probability of finding particle 3 within the interaction volume v_{12} even for free movement along the interface in the absence of an energy barrier can be comparatively small.

A quantitative analysis of this effect as a function of $\dfrac{\delta}{a}$ would be interesting to have.

We shall assume that it is this second factor that is responsible for the formation of linear aggregates in triplets and quadruplets. Furthermore we shall assume that every deaggregation process is an equilibrium one with a probability β. Then the variation in the number of bonds n which are formed in time is given by

$$\frac{dn}{dt} = -\beta n + \frac{1}{2}\alpha z_k^2 \tag{3.45}$$

z_k is the number of species of kind k that are able to form kinetic units with either singlets or linear aggregates. The factor of $1/2$ in the second term takes into account the fact that a contact between particles i and j is identical to that between j and i. The relationship between numbers n, n_k and the total number of singlet particles, z_0, in the formation of linear aggregates is given by

$$n = z_0 - z_k \tag{3.46}$$

Furthermore

$$n = \sum_{i=1}^{z_k} (i - 1)\, z_i \tag{3.47}$$

and

$$z_0 = \sum_{i-1}^{z_k} i \cdot z_i \tag{3.48}$$

If we substitute z_k from equation (3.46) into (3.45), we obtain a simple differential equation for the time evolution of the number of bonds

$$\frac{dn}{dt} = -\beta n + \frac{1}{2}\alpha_{col}\alpha_{ef}(z_0 - n)^2 \tag{3.49}$$

The solution of this equation for the initial stages of coagulation requires

$$n|_{t=0} = 0 \tag{3.50}$$

and

$$\frac{1}{z_0^2}\frac{dn}{dt}\bigg|_{t=0} = \frac{1}{2}\alpha \tag{3.51}$$

Let us now express α and β in terms of parameters that characterize the interaction between the particles.

If the first term in equation (3.49) representing the deaggregation process is still small, i.e., at low levels of aggregation, then, according to von Smoluchowski [equ. (2.29)],

$$\alpha_{col} = 8\pi D_1 R$$

and according to equation (3.22)

$$\alpha_{ef} = \frac{1}{W}$$

From these relationships, together with equations (2.5) and (2.6), the following expression is obtained:

$$\alpha_{col}\alpha_{ef} \equiv \alpha_{exp} = \frac{8}{3}\frac{k_BT}{\eta W} \tag{3.52}$$

For the limiting condition (3.43), the following approximated equation [15] is obtained:

$$W \simeq \frac{1}{2}\frac{\delta_1}{\alpha} \exp\left(\frac{V_{max}}{k_BT}\right) \tag{3.53}$$

where V_{max} represents the height of the energy barrier opposing coagulation and δ_1 represents the effective thickness of this barrier. Since there is no analytical solution for the calculation of an energy barrier, we approximated this by the rectangular potential shown in Figure 3.5. Under these conditions the coefficient α may be determined for the

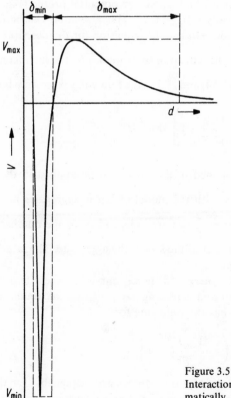

Figure 3.5
Interaction energy—distance curve shown schematically.

early stages of coagulation from the slope of n—t curves. If with increasing number of linear aggregates the deaggregation probability increases, then equilibrium may be established — e.g., the number of bonds becomes constant and

$$\frac{dn}{dt} = 0$$

From equation (3.49) it follows that

$$n_{equ.} = \frac{1}{2} \frac{\alpha}{\beta} (z_0 - n_{equ.})^2 \tag{3.54}$$

were $n_{equ.}$ represents the number of bonds in the equilibrium state.

The equilibrium number of bonds for doublets, at small concentration of singlet particles, is determined by statistical mechanics [16] through the second virial coefficient. This was first considered in reference [13]:

$$n_{equ.} = \frac{z_0^2}{2} \int\limits_{2a}^{2a+\delta_{ij}} \left[\exp\left(-\frac{V}{k_B T}\right) - 1 \right] 4\pi r^2 \, dr \tag{3.55}$$

where δ_{ij} represents the thickness of the rectangular potential profile (Fig. 3.5).

Equation (3.55) is applicably only if $n_{equ.} \ll z_0$. This condition can be fulfilled if z_0 is small enough. However, one does not use this equation for the description of equilibrium in general but one can use this equation to express $\frac{\alpha}{\beta}$ by the interaction energy.

If one compares the equations (3.54) and (3.55), the following expression is obtained:

$$\frac{\alpha}{\beta} = \int\limits_{2a}^{2a+\delta} \left[\exp\left(-\frac{V}{k_B T}\right) - 1 \right] 4\pi r^2 \, dr \tag{3.56}$$

After all the equation for $\frac{\alpha}{\beta}$, obtained in this way, can be applied in the generalized shape [equ. (3.54)] in accordance with the model of linear aggregates in which it was assumed that α and β and therefore $\frac{\alpha}{\beta}$ too are independent of the number of particles within the aggregate and independent of the size of the aggregate. The same assumption was recently made by Muller [88].

If we model the interaction energy—distance curve between two particles by a rectangular potential minimum and a rectangular potential maximum, as shown in Figure 3.5, then equation (3.56) has the following form:

$$\frac{\alpha}{\beta} = 16\pi a^2 \delta_2 \left[\exp\left(-\frac{|V_{min}|}{k_B T}\right) - 1 \right] \tag{3.57}$$

If we assume that only electrostatic repulsion forces are responsible for the repulsion then the characteristic distance in the model (Fig. 3.5) is determined by the thickness of the electrical double layer ($\varkappa^{-1}$). In that case,

$$W = \frac{1}{2\varkappa a} \exp\left(\frac{V_{max}}{k_B T}\right) \tag{3.58}$$

and

$$\frac{\alpha}{\beta} = \frac{16\pi a^2}{\varkappa}\left[\exp\left(-\frac{|V_{min}|}{k_B T}\right) - 1\right] \tag{3.59}$$

Either by determining the equilibrium number of bonds $n_{equ.}$, or from the experimental curves $n(t)$ at two different times t_1, t_2 ... t_i one obtains with equation (3.49) a system of two equations from which α and β and $\frac{\alpha}{\beta}$, respectively, are calculated.

If we introduce the dimensionless number of bonds

$$\tilde{n} = \frac{n_i}{z_0} \tag{3.60}$$

Then α equals

$$\alpha = \frac{2}{z_0} \frac{d\tilde{n}}{dt} \tag{3.61}$$

The dimensionless particle concentration was introduced for the sake of simplicity. If we know, in a given volume of the dispersion, the number of aggregates of any size, then use of the above equations leads to the determination of α and $\frac{\alpha}{\beta}$. From equations (3.52), (3.53), and (3.59) the following parameters are obtained:

$$A = \delta_2\left[\exp\left(-\frac{V_{min}}{k_B T}\right) - 1\right] = \frac{\alpha}{\beta 16\pi a^2} \tag{3.62}$$

and

$$B = \delta_1 \exp\left(\frac{V_{max}}{k_B T}\right) = \frac{16 k_B T a}{3\eta\alpha} = \frac{4 k_s a}{\alpha} \tag{3.63}$$

If we assume, as a first approximation $\delta_1 = \delta_2 = \varkappa^{-1}$, then from the values A and B the primary minimum energy and energy barrier may both be obtained, from the following equations:

$$\frac{V_{min}}{k_B T} = -\ln\left(\varkappa A + 1\right) \tag{3.64}$$

and

$$\frac{V_{max}}{k_B T} = \ln\left(\varkappa B\right) \tag{3.65}$$

The formation of linear aggregates has also been considered in the case of small particles, i.e., for $\varkappa a \approx 1$ in references [17, 18, 19]. For this condition the approach of a particle to a doublet is shown schematically in Figure 3.4b. In this case the probability of a third particle forming a simultaneous contact with both the particles in a doublet is greatly enhanced, compared to the case $\varkappa a \gg 1$.

However, under the condition of slow coagulation, in the presence of an energy barrier, the formation of linear aggregates is still preferred. However, with increasing chain length of the linear aggregate, the probability of "side contacts" also increases:

$$W_E = W_s(L) \tag{3.66}$$

where W_E is the probability of "end contacts" in a linear aggregate, W_s is the probability of a side contact, and L is the chain length. Because there are two "ends" to a chain the probability W_E has to be multiplied by 2. Therefore the total coagulation probability of a single particle with a linear aggregate is given by the sum of both probabilities, i.e.,

$$W = 2W_E + W_s \tag{3.67}$$

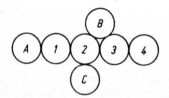

Figure 3.6
Coagulation of single particles with a linear aggregate at different orientations shown schematically.

Various possibilities for coagulation of singles with a linear aggregate are shown schematically in Figure 3.6. The difference in coagulation probability at position A or B has already been considered. We now consider the formation of linear aggregates either at position A (unbranched) or at position C (branched). These two kinds of coagulation are important in the formation of coagulation networks. In analyzing the formation of linear aggregates two factors have to be considered; firstly, the interaction energy between a singlet particle and the linear aggregate, and secondly the mutual diffusion of the singlet and the linear aggregate.

The interaction energy at position A in Figure 3.6 is approximately equal to the interaction between two singlet particles. However, in position C the total interaction energy depends, not only on the pair—pair interaction between particle C and 2, but also on the interaction energy between C and particles 1 and 3. From this simple qualitative consideration, it already follows that the maximum energy at position C is higher than at position A and therefore the aggregation probability is lower. On the other hand, owing to the interaction between C and particles 1, 2, and 3 the deaggregation probability is lower than in position A, since we have a deeper primary minimum at this position.

From these considerations it follows that unbranched linear aggregates should be obtained preferentially under these conditions.

However, as the number of particles in linear, unbranched chains increases, the number of coagulation sites in the chain also increases.

The result is increasing aggregation probability at a "side" position. At a certain chain length both the aggregation probabilities considered should become equal, in line with equation (3.66)

$$W_E = W_s(L)$$

Let us now model the mutual diffusion of a sphere and a linear aggregate as the diffusion between a sphere and a cylinder of radius a and length L, $L = na$.

We have to distinguish between the translatory diffusion along the axis and across the axis and we also have taken into consideration the rotary diffusion.

According to Müller (ref. [22] in chapter 2 — see section 2.2.6.) an equivalent radius a^* is introduced for an anisotropic particle, i.e., a sphere having the same mobility as the anisotropic aggregate.

The equivalent radius of linear particles, for diffusion along the axis is given approximately by

$$a^* \approx \frac{2}{3} \frac{b}{\ln\left(2\frac{b}{c}\right)} \tag{3.68}$$

and that for diffusion across the axis by

$$a^* \approx \frac{4}{3} \frac{b}{\ln\left(2\frac{b}{c}\right)} \tag{3.69}$$

For a linear aggregate of spherical particles, $b = na = L$ and $c = a$.

The rotary diffusion may also be expressed in terms of an equivalent radius [20]

$$a^*_{\text{rotation}} \approx b\left[3\ln\left(2\frac{b}{c}\right)\right]^{-1/3} \tag{3.70}$$

The total diffusion coefficient is the sum of the translatory diffusion D^T and the rotary diffusion $L^2 D^R$

$$D = D^T + L^2 D^R \tag{3.71}$$

where D^T is the translatory diffusion coefficient, and D^R is the rotary diffusion coefficient.

In Table 3.1 the relative diffusion coefficient of two spheres, $D_{s,s}$ is compared

Table 3.1

Ratio of the relative diffusion coefficient of a spherical and ellipsoidal particle having different half-axis ($D_{e,s}$) and two spherical particles ($D_{s,s}$)

b/a	5	10	20	50
$\dfrac{D_{e,s}}{D_{s,s}}$	1.25	1	0.8	0.5

with the relative diffusion coefficient of a sphere and ellipsoidal particle $D_{e,s}$, having different values for their half-axis, $\dfrac{b}{c}$. In Table 3.2 the relative diffusion coefficients of two ellipsoidal particles, $D_{e,e}$, are compared with the relative diffusion coefficient of two spheres. It follows from both tables that the influence of the shape on the relative diffusion coefficient is not very large. We should recall that the approximation of an aggregate of spherical particles by an ellipsoid or cylinder is only applicable for small aggregation numbers. With higher aggregate numbers the flexibility of the particle chain cannot be neglected and one would expect a collapse of the particle chain, comparable with coil formation in macromolecules.

Table 3.2

Ratio of the relative diffusion coefficient of two ellipsoidal particles and two spherical particles

b_1/b_2	$\dfrac{10}{5}$	$\dfrac{20}{5}$	$\dfrac{20}{10}$
$\dfrac{D_{e,e}}{D_{s,s}}$	1.25	1.05	0.8

Now let us consider the flux of spherical particles, either to the end region of the cylinder (particle A in Fig. 3.6), or to the side region (particle B or C), under the influence of the interaction energy between the coagulating particles.

According to equation (2.37)

$$I = \frac{4\pi D_{12} z_0}{\displaystyle\int_{2a}^{\infty} \exp \frac{V(r)}{k_B T} \frac{dr}{r^2}} \qquad (2.37)$$

But we have to remember that there are two end regions, therefore one has to multiply the right-hand side of the above equation by a factor of 2. For the side region the same equation (2.39) is valid, but with the restriction that the interaction energy cannot be approximated by the interaction energy of two spheres. Two different approximations may be proposed for the interaction of particle C with the linear aggregate (Fig. 3.6). Firstly, one may consider the interaction of particle C separately with the three particles 1, 2, 3. Then the total interaction energy, V^{123}, is given by

$$V^{123} = V(r) + 2V(r)\sqrt{r^2 + 4a^2} \qquad (3.72)$$

A second approach would be to consider the linear aggregate as a cylinder. The interaction energy between a sphere and a cylinder has been described in ref. [18].

The attraction energy between a spherical particle and a particle of any shape, with volume v_1, is given by the following equation:

$$V_A = -\frac{4a^3 A}{3\pi} \iiint_{v_1} \frac{dv_1}{\varrho^6} \tag{3.73}$$

If the particle is a cylinder of radius a and of infinite length, expression (3.73) can be written as

$$(V_A)_{SC} = -\frac{4Aa^3}{3\pi} \int_{-a}^{+a} dx \int_{-\infty}^{+\infty} dz \int_{-\sqrt{a^2-x^2}}^{+\sqrt{a^2-x^2}} \frac{dy}{[(d + 2a + y)^2 + x^2 + z^2 - a^2]^3} \tag{3.74}$$

After integration and by introducing the relative (dimensionless) distance, $s = \dfrac{r}{a}$, the following equation is obtained for the van der Waals attraction between a sphere and a cylinder:

$$(V_A)_{SC} = \frac{A}{3}\left(\frac{1}{s^{3/2}(s-2)} + \frac{0.833}{s^{3/2}} + \frac{1.25}{s^{1/2}} + 0.442 \ln \frac{u^{1/2} - 1.41}{s^{1/2} + 1.41}\right) \tag{3.75}$$

and for the electrostatic repulsion,

$$V_{el} = \left[\frac{4\pi\varepsilon\varepsilon_0 a\psi_0^2}{k_B T}\right] \cdot \frac{1}{K_0(\varkappa a)} \cdot \frac{\exp\left[-\varkappa a(u-1)\right]}{(\varkappa a s)^{1/2}}$$

$$\times \left(1.253 - \frac{0.1566}{\varkappa a s} + \frac{0.08758}{(\varkappa a s)^2}\right) \tag{3.76}$$

where

$$K_0(\varkappa a) = \frac{\exp(-\varkappa a)}{\sqrt{\varkappa a}}\left(1.253 - \frac{0.1566}{\varkappa a} + \frac{0.08758}{(\varkappa a s)^2}\right) \tag{3.77}$$

The total interaction energy between a sphere and a linear aggregate can now be calculated using either equation (3.72) or equations (3.75) and (3.76).

The diffusional flux to the side region is given by

$$I \cdot L = \frac{4\pi D_{12} z_0 L}{\displaystyle\int_2^{\infty} \exp\left(\frac{V(s)}{k_B T}\right) \frac{ds}{s}} \tag{3.78}$$

If we now consider equilibrium, so that the probability of coagulation at the end of linear aggregates equals the probability of coagulation at the side region, a critical length, L_{cr}, can be calculated using equations (2.37) and (3.78), i.e.,

$$L_{cr} = 2a \frac{\displaystyle\int_2^\infty \exp\left(\frac{V(s)}{k_B T}\right) \frac{ds}{s}}{\displaystyle\int_2^\infty \exp\left(\frac{V(s)}{k_B T}\right) \frac{ds}{s^2}} \tag{3.79}$$

If we now compare the values for the critical length (Table 3.3) calculated either by the sphere—sphere interaction L_{ss} or sphere—cylinder interaction L_{sc}, then for all the combinations of van der Waals interaction, Stern potential, and particle size considered, $L_{ss} > L_{sc}$.

Table 3.3
Chain length L, calculated in different approximations

	$A = 10^{-19}$ J, $a = 7,5$ nm						$A = 10^{-19}$ J, $\psi_0 = 25.6$ mV								$a = 10$ nm, $\psi_0 = 25.6$ mV									
	ψ_0 in mV						a in nm								A in J									
	25.6			35			5		10			15			$1.0 \cdot 10^{-19}$			$0.5 \cdot 10^{-19}$			$0.1 \cdot 10^{-19}$			
x^{-1} in nm	2.5	5	7.5	2.5	5	7.5	5	10	5	7.5	10	5	7.5	10	2.5	5	7.5	2.5	5	7.5	2	2.5	4	5
L_{ss}	2	6	12	3	21	106	7	16	5	12	27	3	8	24	2	5	12	5	6	16	2	3	5	8
L_{sc}	2	3	4	3	7	13	2	6	2	4	7	2	4	7	2	3	4	2	4	7	5	6	10	12

As expected from the DLVO theory, the length L increases with the double layer thickness and thus increasing height of energy barrier, or with increasing surface potential at constant $\varkappa$. An increase in the Hamaker constant reduces the size of a linear aggregate. If we analyze the influence of particle radius on the chain length, at constant $\varkappa$, then the chain length is independent of a for the cylinder approximation and increases irregularly with a for the sphere approximation.

Since the interaction potential in the DLVO theory is independent of the particle concentration, the average length of the chains formed during slow oriented irreversible coagulation should be independent of the particle concentration at least at low particle concentrations.

3.5.   Rate Effect in Particle Collisions

The calculations carried out above of the interaction between coagulating (approaching) particles were mostly made with the assumption that the overlapping electrical double layers are always at equilibrium at all separations of the particles. As was shown by Overbeek [84], and later by Lyklema [85], this is not necessarily correct, since the adjustment of equilibrium takes time. This relaxation time may be significantly longer than the collision (coagulation) time.

Calculation of the electrostatic interaction at constant surface charge (section 3.4.) is based on the assumption that no discharge of the surface did occur during the collision (unrelaxed double layer). If constant potential is assumed during the collision (fully relaxed double layer), then equilibrium adsorption is always established in the double layer. In the latter case the relaxation time depends on the diffusion of the ions through the double layer and on the interaction distance, $2\varkappa^{-1}$. In that case one obtains for the relaxation time

$$t_{\text{rel}} = \frac{(2\varkappa^{-1})^2}{2D_{\text{ion}}} \sim 10^{-8} \text{ s} \tag{3.80}$$

The time of a Brownian collision, that is the time for a particle to diffuse a distance of the order of the thickness of the double layer, is defined by

$$t_{\text{Brown}} = \frac{(2\varkappa^{-1})^2}{2D_1} \tag{3.81}$$

If we substitute for D_1 using equation (2.5), we obtain the following equation:

$$t_{\text{Brown}} = \frac{12\pi\eta a}{\varkappa^2 k_{\text{B}}T} \tag{3.82}$$

For particles with radii of 100 nm and for electrolyte concentrations between 10^{-3} and 10^{-1} mol $\cdot$ dm^{-3}, with $\varkappa^{-1}$ varying from 10 to 1 nm, the time for the mutual diffusion of two particles is given by

$$t_{\text{Brown}} = 10^{-7}-10^{-5} \text{ s} \tag{3.83}$$

However, as was mentioned in section 2.2.2., the approach of particles is slowed down by the so-called "hydrodynamic interaction". Particle approach will take place only when van der Waals forces are present. The hydrodynamic interaction decreases the time for mutual diffusion by an order of magnitude.

The relaxation of the nondiffuse part of the double layer is determined by the exchange current density (i_0). This parameter can vary drastically for different systems, and it was assumed by Overbeek that

$$t_{\text{charge}} \sim \frac{\sigma_0}{i_0} = \frac{10^{-6} \text{ As} \cdot \text{cm}^{-2}}{(10^{-10} - 1) \text{ A} \cdot \text{cm}^{-2}} = 10^{-6}-10^4 \text{ s} \tag{3.84}$$

The relaxation time of this process may be much faster or much slower than the time scale of a Brownian collision.

The lateral diffusion in the Stern layer is determined by the following equation

$$t_{\text{Stern}} = \frac{a}{2D_i^s} \tag{3.85}$$

D_i^s is the surface diffusion coefficient of the ions in the Stern layer.

It is reasonable to suppose that the surface diffusion coefficient of ions is less than the volume diffusion coefficient and therefore the relaxation of the Stern layer should be slower then the relaxation of the diffuse double layer.

3.6.   Methods of Investigating Coagulation Kinetics

3.6.1.   Ultramicroscopy

Small particles which are not detectable in bright-field microscopy may become visible under dark field illumination, because of their diffraction image. The diffraction image of a particle is always circular, independent of the particle shape, but anisotropy of shape may be detected by the variation of the scattered light intensity. This results from the different orientations of the particles in the light beam caused by their Brownian motion. The lower limit of detection of small particles depends on the ratio of their refractive index to that of the dispersion medium. In aqueous dispersions of lyophobic particles, which in general have relatively large refractive index ratios, particles greater than about 6 nm radius may be detected, but with aqueous dispersions of lyophilic particles the lower limit is about 40 nm.

Microscopes in which particles are visible by virtue of their scattered light are called ultramicroscopes. The first ultramicroscope was described by Siedentopf and Zsigmondy [21]. Its operation is shown schematically in Figure 3.7. A light beam of high intensity — for example from a carbon arc lamp — is focused by the telescope lens onto the bilateral slit and a second telescope lens images this slit in the image plane of the condensor lens of the microscope. The glass cell requires two windows arranged perpendicularly to each other. The observation volume in which the particles are counted depends on the depth of focus and on the ocular diaphragm. With this arrangement artefacts may occur as a result of convection within the cell, since the incident light beam (of high intensity) is unidirectional. In place of the slit ultra-

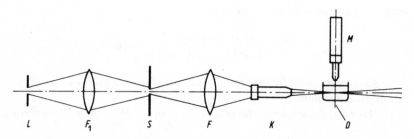

Figure 3.7
Scheme of the slit ultramiscroscope according to Zsigmondy and Siedentopf [21].
L, light source (carbon arc lamp); F_1, telescope lens; S, bilateral slit; K, condensor lens; F, telescope lense with focal length of 80 mm; M, microscope; D, colloidal dispersion.

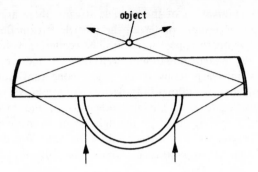

object

Figure 3.8
Principle of cardioid condensor.

microscope a cardiod condensor (Fig. 3.8) ultramicroscope can be used. The incident light beam is now reflected by two spherical mirrors, one of which is convex and the other concave. In this way, only the light scattered by the particles is seen in the microscope. However, the advantage of a higher scattered light intensity is offset by the lower observation volume. Therefore, this kind of ultramicroscope is not suitable for kinetic experiments.

The number of particles may be determined by visual counting of the light flashes within the scattering volume. However, this may lead to counts which are too high because the light scattered by the particles may illuminate particles outside the scattering volume itself. This effect increases with the size of the particles or aggregates. Nevertheless this method was used with some success in the 1920s and 1930s to investigate rapid and slow coagulation. Some of the results of these early experiments are summarized in section 3.6. Rapid coagulation was often investigated by "fixing" the number of particles at various times by adding macromolecular solutions, e.g., gelatin to stop the coagulation process.

This method is very time consuming, inconvenient, and not very accurate. A much better method, instead of counting the number of particles in a fixed volume of the dispersion, is to count the number of particles traversing a fixed field of view as the dispersion flows past.

Some progress in this direction was made in the investigation of coagulation processes with the development of the streaming ultramicroscope. This method was developed in the 1940s and 1950s for particle counting in aerosols, firstly by Derjaguin and Vlasenko [22—24] and then by Gucker and O'Konsky [25—27].

Later this method was also applied to colloidal dispersions in liquids. An especially important advantage of the flow method is the considerably smaller time needed to measure low particle concentrations, compared with the classical ultramicroscope method of Siedentopf and Zsigmondy. In addition the counting volume in the dispersion is better defined and a wider range of particle sizes ($0.05—10\ \mu m$) can be investigated. The size of the counting volume limits the particle concentration, since only one particle is allowed to be in the scattering volume at any time. Derjaguin proposed a counting volume of about $3 \cdot 10^{-11}\ cm^3$.

With regard to the actual observation of the particle stream, there are two kinds of streaming ultramicroscopes. Derjaguin preferred the particle stream flow to be in the direction of the axis of the microscope (Fig. 3.9). The scattering volume is then dependent on the size of the ocular diaphragm well defined and on the depth over which the particles are in sharp focus in the microscope (not well defined). Therefore, this kind of ultramicroscope allows only particle countings per unit volume but no measurements of aggregate size. On the other hand, the image of the particles remains fixed during its passage through the light beam. In this configuration the scattered light energy is greater, and, hence, the size limit for detection of smaller particles is lower than in apparatus in which the forward scattering is measured.

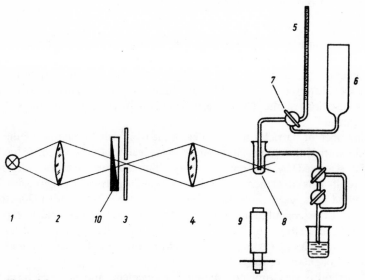

Figure 3.9
Scheme of flow ultramiscroscope according to Derjaguin and Vlasenko [24].
1, mercury lamp; *2*, lens (focal length 60 mm); *3*, diaphragm; *4*, microscope objective with aperture 0.30; *5*, microburet; *6*, buret; *7*, three-way tap; *8*, cell; *9*, microscope; *10*, photometric wedge.

In the second kind of streaming ultramicroscope the direction of observation is perpendicular to the particle stream (Fig. 3.10).

A Derjaguin apparatus is shown schematically in Figure 3.9. The light from a mercury lamp (*1*) is focused by the lens (*2*) (focal length 6 cm) on the diaphragm (*3*). This "pointlike" source is then focused with the lens (*4*) into the particle stream. The streaming cell consists of two cylindrical burettes (*5*, *6*), connected by the three-way tap (*7*). Using the microburette (*5*) the volume of the dispersion is measured. The flow rate used was about $0.03-0.1$ cm^3 · min^{-1}. The particles were counted visually with the help of the microscope (*9*). This equipment allows measurement up to particle concentrations of about 10^{10} particles per cm^3. By introducing a photometric wedge (*10*) the intensity of the incident light beam could be varied and, therefore, also the lower limit of detectibility of particle size. Particle flow along the direction of observation was used in the apparatus of McFadayen and Smith [28]. However, in most streaming ultramicroscopes the direction of observation is perpendicular to the direction of flow [25—27, 29—31, 91]. In reference [35] the direction of observation was at an angle of 20 or 30 degrees.

These kinds of ultramicroscopes can be used, in principle, for the direct measurement of the scattered light intensity from single particles or aggregates [31].

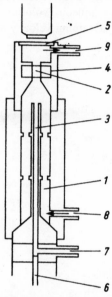

Figure 3.10
Flow ultramicroscopic cell according to McFadayen and Smith [28].
1, outer capillary (diameter 6 mm); *2*, reduced capillary (diameter 2 mm); *3*, inner capillary; *4*, window for the inlet of the laser beam; *5*, observation window; *6*, micrometer syringe; *7*, inlet for filling the micrometer buret; *8*, inlet for the focusing water stream; *9*, outlet.

The next stage in the development of ultramicroscopes was the construction of a cell with hydrodynamic focusing of the particle stream [28]. This arrangement overcomes the chief disadvantage of the earlier ultramicroscopes, namely, blockages in the capillary. The principle of this method is shown in Figure 3.10. The particles are injected into a stream of flowing, carefully filtered water, optically clean. The particle stream flows in the direction of the axis of the microscope. The streaming velocity of the pure water is slightly greater than the velocity of the dispersion. In the conical part of the capillary (*1*) the diameter is reduced from 6 to 2 mm. Hence the flow velocity is enhanced, before the particles reach the illuminated zone and are counted. The upper section of the cell is constructed from black Perspex glass to adsorb the laser beam and hence avoid reflection. The water stream is supplied by gravity feed, the flow rate being controlled by a valve (Gapmeter flow gauge). The dispersion is injected, at constant flow rate, by means of a syringe coupled to a synchronous motor. The light source is a 8-mW He—Ne-laser. A photomultiplier is used as the detector.

The cell we have used [31, 34] is similar to that of Smith et al. except that the head of the cell was changed, as shown in Figure 3.11. The dispersion flows through a stainless steel capillary (1) of 1 mm diameter, driven by a micrometer screw coupled to a synchronous motor whose speed can be changed. The maximum injection rate is $1.6 \cdot 10^{-5}$ cm$^3 \cdot$ s^{-1} and the average shear gradient is 0.1 s^{-1}. At the end of this capillary a smaller stainless steel capillary of 80 µm diameter is fixed. From this capillary the dispersion flows into the stream of the focusing aqueous solution having the same electrolyte concentration as the dispersion. The streaming velocity at this point is about 0.3 cm $\cdot$ s^{-1}.

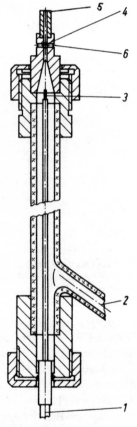

Figure 3.11
Schematic of the flow ultramicroscopic cell according to Gedan and co-workers [31, 34].

1, Stainless steel capillary of 1 mm diameter; *2*, glass capillary of 7.5 mm, reduced in the conical section to 2 mm; *3*, stainless steel capillary of 80 µm diameter; *4*, observation window; *5*, particles outlet; *6*, observation window.

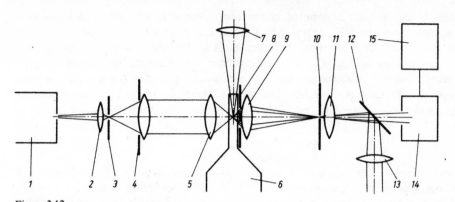

Figure 3.12
Schematic of the apparatus for measuring single particle laser light scattering [31, 34].
1, Argon laser; *2* focusing lens; *3*, pinhole; *4*, lens to parallelize the light beam; *5*, focusing lens; *6*, streaming cell; *7*, microscope; *8*, annular stop; *9*, objective; *10*, diaphragma; *11*, ocular; *12*, beam splitter; *13*, microscope; *14*, photomultiplier; *15*, multi channel analyzer.

The focusing electrolyte solution flows through a glass capillary of 7.5 mm (*2*). The diameter in the conical section is reduced to 2 mm and hence the particle stream to about 20 μm. The focusing electrolyte solution is cleaned by membrane filters. A constant streaming velocity is obtained, by means of a constant level of the focussing liquid in a large volume reservoir. The streaming velocities of the dispersion and of the focusing electrolyte solution are made equal.

The particle stream can be observed by a microscope through window (*4*); (*5*) is the outlet of the streaming liquid. The laser beam is focused into the particle stream through window (*6*); a window on the opposite side is the exit for the primary beam and also the scattered light. It thus follows that in this arrangement the forward scattering is detected.

The complete apparatus is shown schematically in Figure 3.12. Monochromatic light from a 1 W argon-ion laser ($\lambda = 488$ nm) is focused by lens (*2*) on a pinhole of diameter 12 μm (*3*). The focal point of lens (*4*) is located in the center of the pinhole and produces a quasi-divergence-free light beam. With lens (*5*) (focal length ⟩50 mm) the parallel light beam is focused into the center of the particle stream. To obtain a constant intensity profile within the particle stream the lens was mounted at an angle of 5° to the optical axis with the aid of a special calibrating device.

The cell may be calibrated. The scattering volume can be observed with a microscope under 90° to the optical axis.

The optical system to measure the forward scattering consists of an annular stop (*8*), an objective (*9*), a diaphragm (*10*), an ocular (*11*), a beam splitter (*12*), a microscope for observation in the forward direction (*13*), a photomultiplier (*14*), and a multichannel analyzer (*15*). The analyzer selects and counts the impulses according to their intensity.

The scattering volume is defined as the volume formed by the diameter of the light beam and the diameter of the particle stream. Every particle, or aggregate of particles, in this volume is counted and stored in the analyzer according to its size. The size of the scattering volume and the particle concentration are coupled. The higher the particle concentration the lower must be the observation volume, so that only one particle (or aggregate) exists in this volume. For a random distribution of particles, the probability distribution for the number of particles (in a given volume element) is a Poisson distribution. For a coincidence error of 0.5% the following relation is obtained between the maximum particle concentration z_{max} and the scattering volume v:

$$z_{max} = 0.1v^{-1} \tag{3.86}$$

The coincidence error is defined as the ratio of the probabilities that two (or more) particles or only one particle is present in the scattering volume. From this it follows that for a particle concentration of 10^8 particles $\cdot$ cm^{-3} the scattering volume may not exceed 10^{-9} cm^3. Hence if the maximum diameter of the laser beam is 8 μm then the maximum diameter of the particle stream is 20 μm. Besides the low scattering volume, a constant intensity profile of the laser beam is necessary. This was obtained by the defocusing of an aberration-free lens. With this artificial astigmatism a constant intensity was obtained within the laser beam.

To show the high resolution of this equipment, a mixture of three monodisperse polystyrene latices with diameter 0.50 μm, 0.52 μm, and 0.60 μm (from electron microscopy) were analyzed (Fig. 3.13) [35]. The scattered light from every particle or aggregate passing the scattering volume is measured separately and stored in the

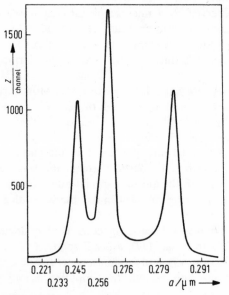

Figure 3.13
Number of impulses z of particles in a mixture of three different latices with diameters 0.50, 0.52, and 0.60 μm, respectively, according to reference [34].

mutichannel analyzer. However the resolution power with respect to the light scattered by aggregates is limited. The reason for this is that the scattered light is uniquely defined only for singlet and doublet particles. An aggregate of three single particles has already two possible orientations as a linear or compact aggregate. The number of possible orientations of an aggregate increases with the number of particles within that aggregate. On the other hand, an exact theory for particle light scattering exists only for spheres and cylinders. An aggregate of two spherical particles may reasonably be approximated by a cylinder. However, this is not reasonable for a linear aggregate of three particles, owing to the flexibility of the aggregate chain. This fact limits the interpretation of the results of single-particle laser light scattering to singlet, doublet, triplet, and quadruplet particles.

The scattered light impulses may be analyzed using the Rayleigh-Debye-Gans (RDG) theory in which aggregates are assumed to be spheres. This is valid for the condition

$$m2a \ll \frac{\lambda}{2\pi} \tag{3.87}$$

where m is the relative refractive coefficient of the particle in the dispersion medium, and λ is the wavelength of the light in water.

The intensity of the scattered light i of a spherical particle is given by

$$i_\vartheta = \frac{(1 + \cos^2 \vartheta) \, \alpha^2 \, 16\pi^4 v}{2r^2 \lambda^4} \, I_0 \tag{3.88}$$

where I_0 is the intensity of the incident light, α is the polarizability, r is the distance of the receiver from the scattering particle, ϑ is the scattering angle, and v is the volume of the scattering particle.

The factor of 2 in the denominator is necessary because each component of the polarized light has half the intensity of the incident light I_0.

The polarizability can be determined from the relative refractive index and the scattering volume v, i.e.,

$$\alpha = \frac{m^2 - 1}{4\pi} \, v \tag{3.89}$$

Substituting equation (3.89) in (3.88), for light polarized perpendicularly to the plane of incidence and for zero scattering angle, equation (3.88) reduces to

$$i_{\vartheta = 0} = \frac{16\pi^4 v^2}{\lambda^4 r^2} \left(\frac{m^2 - 1}{4\pi} \right)^2 I_0 \tag{3.90}$$

For scattering angles $\vartheta > 0$ a form factor, $R(\vartheta, \varphi)$, has to be introduced, i.e.,

$$R(\vartheta, \varphi) = \frac{1}{v} \int e^{i\delta} \, \mathrm{d}v < 0 \tag{3.91}$$

where φ is the orientation of the particles with respect to the incident light beam and $e^{i\delta}$ is the phase factor which refers the scattered waves in a given direction to a common origin of the coordinates.

The boundary conditions for the RDG theory are

$$m - 1 \ll 1$$

and also

$$\frac{2\pi a}{\lambda}(m - 1) \ll 1$$

Recently Smith and co-workers [33] published a new version of their flow ultra-microscope, with which size distributions are also measurable. Their apparatus is shown schematically in Figure 3.14. A feedback mechanism from the photomultiplier to the laser was used to ensure that the dynamic range of the photomultiplier is now exceeded in the formation of aggregates. A 40-mW He—Cd-laser (441 nm) was intensity modulated to 1 MHz using a voltage feed of 0—1 V. A potential divider circuit on the output of the photomultiplier feeds a signal to the laser which is proportional to the light intensity falling on the photocathode, leading to a consequent change in the laser output. With this method an approximately 760-fold increase in the dynamic range was obtained. A light beam of elliptical cross section (30 by 100) nm was obtained using a combination of a cylindrical and a spherical lens. The streaming cell was the same as that shown in Figure 3.10. The height of the scattered light impulse was measured using a photomultiplier and/or an oscilloscope. A mixture of five latices was investigated. The results are shown in Figure 3.15.

Walsh and others [35] have developed a flow ultramicroscope for particle size distri-bution analysis which will count particles in liquids with diameters between 0.1 and 1 μm. The operation principle is similar to those described above, but it is different in respect to the angle of the scattered light. The block diagram of the apparatus is shown in Figure 3.16; two different cells for measuring light scattering at 20° and

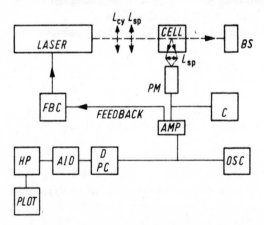

Figure 3.14
Schematic of experimental arrangement according to Cummins et al. [38].
L_{cy}, L_{sp}, cylindrical and spherical lens; BS, beam stop; C, pulse counter; FBC, feedback loop control; D, PC, discriminator and peak catcher; PM, photomultiplier; OSC, oscilloscope, A, D, analog-to-digital converter; HP, calculator; AMP amplifier.

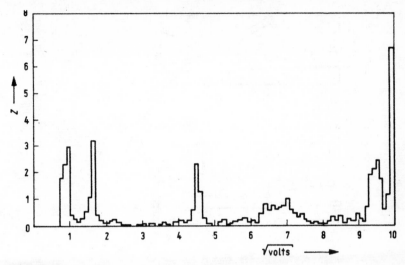

Figure 3.15
Particle size distribution of five different latices with diameters 0.22, 0.5, 1.0, 1.6, and 3.2 μm. The x axis is represented as a relative square root of the measured voltages [36].

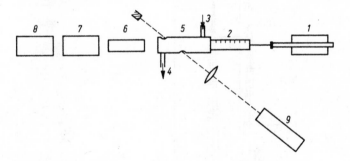

Figure 3.16
Block diagram of the apparatus according to Walsh et al. [35].
1, infuser; 2, syringe; 3, water in; 4, water out; 5, streaming cell; 6, microscope; 7, photomultiplier; 8, pulse height detector and counter; 9, He—Ne laser.

30° are shown in Figure 3.17. Using different kinds of infusers the injection rate was modified. The particle stream is focused hydrodynamically to a diameter of 2 mm. A He—Ne-laser beam (2.5 mW) passes through a 1-mm channel at 30° to the axis. The laser is focused with a convex lens ($f = 10$ cm) to a diameter of about 100 μm. The focusing water stream was cleaned using membrane filters. The flow rate was 9 cm³ · min⁻¹. The impulses were counted according to their height. A comparison of the size distribution data obtained from electron microscopy and ultramicroscopy is shown in Figure 3.18. The electron micrographs represent data from about 1000 particles, the ultramicroscopic data from 25000 particles. The loss of small numbers of particles in

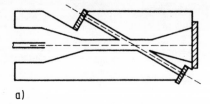

a)

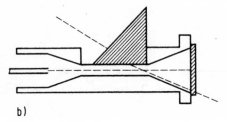

b)

Figure 3.17
Scattering cells for different angles (a) 30°, (b) 20°.

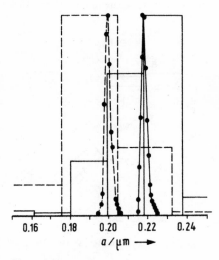

Figure 3.18
Comparison of electron microscope and flow ultramicroscope size distributions. Histograms refer to electron microscope results, the points to flow ultramicroscopy [35].

outlying populations is inevitable in flow ultramicroscopy, as they may disappear into the background. The authors also compared results using this "optical" counter with a Coulter counter (see section 3.6.2.). The results from the two methods were 8.6 and $8.5 \cdot 10^6$ particles $\cdot$ cm^{-3}, respectively.

The use of an ultramicroscope particle counter in colloidal dispersions in liquids, including aerosol formation has also been proposed [36, 37]. However, it is uncertain whether this method can be applied to aggregates because during the aerosol formation deaggregation (or even aggregation) may occur.

Recently an optical pulse particle size analyzer was described by Bowen et al. [87].

3.6.2. Coulter Counter

Particle counting with the Coulter counter is based on the change of the electrical resistance when a nonconducting particle in an electrical conducting medium passes through a short capillary, whose diameter may vary from 20 to 400 µm. At each orifice of the capillary an electrode is located. When a dilute dispersion passes through the capillary each particle produces a voltage or current impulse if their conductivity is different from the conductivity of the background electrolyte solution. The height of the impulse is dependent on the volume of the particles and also on their velocity. The number of impulses and their height can be stored using an impulse analyzer. The change in the resistance (ΔR) during the passage of a particle, radius a, through the capillary, radius r, is given by the following equation:

$$\Delta R = \frac{\varrho_0}{\pi r} \left\{ \frac{\cot \left[\frac{a}{r} \left(1 - \left[\frac{a}{r} \right]^2 \right)^{1/2} \right]}{\left(1 - \left[\frac{a}{r} \right]^2 \right)^{1/2}} - \frac{a}{r} \right\} \tag{3.92}$$

where ϱ_0 is the specific resistance of the dispersion medium.

The sensitivity of the apparatus is very dependent on the ratio of the particle radius to the capillary radius. If this ratio is less than 0.02, the method becomes rather insensitive.

Usually the lower limit of the capillary radius is about 10 µm owing to possible blockage of the capillary with particles. This limits the application of this method to particles of ~ 0.2 µm or greater (e.g., emulsion droplets). The sensitivity of the method is also dependent on the difference between the conductivity of the particles and dispersion medium. Therefore, electrolyte has to be added, which may reduce the stability of the dispersion. The Coulter counter monitors particle volumes, and, hence, provided the original dispersion is monodisperse, singlets, doublets, triplets, and even quadruplets can be detected and counted [38—40].

3.6.3. Measurement of the Volume Scattering in Colloidal Dispersion

There are two principal ways of monitoring light scattering in colloidal dispersions. The first method is to measure the scattered light directly at some fixed angle to the incident beam. This is the counterpart of the single particle scattering described in section 3.6.1. The second method is to measure the turbidity, that is the reduction in intensity of the transmitted light due to the scattering of the colloidal particles.

There are many types of light scattering apparatus available commercially, detailed descriptions will not be given here. The intensity of the scattered light from a single particle is given by equation (3.88). For z particles, and for an isodisperse system, the following equation holds:

$$i_\vartheta = z \frac{(1 + \cos^2 \vartheta)\, \alpha^2 16\pi^4 v}{2r^2 \lambda^4} I_0 \tag{3.93}$$

The turbidity τ is defined as follows:

$$\tau = \frac{1}{l} \ln \frac{I_0}{I} \tag{3.94}$$

where l is the scattering path length and I is the intensity of the transmitted light.

The turbidity and the optical density D are related by

$$\tau = 2.303 \frac{D}{l} \tag{3.95}$$

The turbidity in the Rayleigh region $\left(a < \dfrac{\lambda}{5}\right)$ is given by

$$\tau = Cv_0^2 z_0 \tag{3.96}$$

where C is an optical constant and v_0 is the volume of the particle.

The turbidity $\tau(t)$ as a function of time during coagulation is given by a combination of the equation (3.96) with equation (3.7). At the early stages of coagulation, where

$$z_2 \ll z_1$$
$$\tau(t) = Cv_0^2 z_0 (1 + k_s z_0 t)^{-1} \tag{3.97}$$

Using this equation the Smoluchowski coagulation rate constant can be determined for small particles.

For larger particles Mie theory has to be applied. This theory is valid for spherical particles of all sizes and refractive indices. Mie theory considers the intensity distribution of scattered light with angle for a vertically polarized primary beam. The forward scattering is somewhat larger than the backscattering. The scattering intensity is (except at zero scattering angle) somewhat lower than for Rayleigh scattering. The intensity profiles for particles with $2a > \lambda$ show maxima and minima. Their position and magnitude is characteristic for a given particle size. Calculations according to Mie theory require extensive computations although tables of data are available [46].

The turbidity, for an initially mondisperse system, is given by the cross section of a single particle (πa_1^2), the total scattering coefficient, C_1, and the number of particles, z_0, i.e.,

$$\tau_0 = \pi a_1^2 C_1 z_0 \tag{3.98}$$

During coagulation higher aggregates are formed and therefore the turbidity changes. Utilizing the kinetic theory of von Smoluchowski, the following equation is obtained:

$$\tau_t = \pi a_1^2 C_1 z_1 + \pi a_2^2 C_2 z_2 + \pi a_3^2 C_3 z_3 \ldots \tag{3.99}$$

If we consider the early stages of coagulation, where only doublets are formed, and if we assume coalescence ($a_2 = 1.26\,a_1$), then

$$\tau = \pi a_1^2 C_1 n_1 + \pi a_2^2 C_2 n_2 = \pi a_1^2 (C_1 + 1.59 C_2) \tag{3.100}$$

The following equation (3.101) is obtained for $t \lesssim 1/10\, T_{ag}$ where T_{ag} is defined [see equ. (3.9)] as the time in which half the singlet particles are coagulated:

$$\frac{d\tau}{dt} = \pi a_1^2 C_1 \frac{dz_1}{dt} + \pi a_2^2 C_2 \frac{dz_2}{dt} = \pi a_1^2 \left(C_1 \frac{dz_1}{dt} + 1.59 C_2 \frac{dz_2}{dt} \right) \tag{3.101}$$

According to equations (3.3) and (3.4),

$$\left. \frac{dz_1}{dt} \right|_{t \to 0} = -k_{11} z_1^2 \quad \text{and} \quad \left. \frac{dz_2}{dt} \right|_{t \to 0} = \frac{k_{11}}{2} z_1^2$$

and therefore

$$\left. \frac{d\tau}{dt} \right|_{t \to 0} = k_{11} \pi z_1^2 \left(\frac{a_2 C_2}{2} - a_1 C_1 \right) = k_{11} \pi z_1^2 (0.795 C_2 - C_1) \tag{3.102}$$

The coagulation rate constant for doublet formation can be calculated using this equation.

In measuring turbidity the angle of illumination and the angle of acceptance should be low and less than 1° to prevent both primary and multiple scattered light from reaching the photomultiplier. Suitable instruments have been described by many authors [43—45]. Cuvettes of black material with the exception of the inlet and outlet window should be used in order to minimize reflection of the cell wall. The mixing dead time should ideally be a few milliseconds. In most coagulation experiments longer times of observation than this are necessary in order to achieve a change of transmission of at most 1%, which is in general the level of noise in the signal. On the other hand, as mentioned above, the observation time should not be longer than $\sim \dfrac{T_{ag}}{10}$ for equation (3.102) to be valid.

For light transmission experiments the lens—pinhole system is recommended [47, 48]. The apparatus used in reference [48] is shown schematically in Figure 3.19.

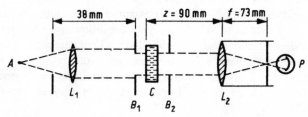

Figure 3.19
Scheme of the apparatus of Melik [48] to measure turbidity of coagulating dispersions.

Light from a monochromator A is focused with lens L_1 into a nearly parallel beam. A beam stop B_1 determines the diameter of the light beam, which then passes through the cuvette C. The beam stop, B_2, eliminates secondary scattered light and lens L_2 focuses the beam at the pinhole which is located at the focal length of this lens. The pinhole must be large enough to allow all the light in the focused beam to pass through and strike the cathode of the phototube. The value of the half-angle, ϑ, of the detector is equal to $\tan^{-1}\left(\dfrac{a}{f}\right)$, where a is the radius of the pinhole and f the focal length of lens L_2. This lens—pinhole system practically excludes all stray light from the phototube.

Since a point source is physically impossible, a perfectly collimated incident beam is not feasible; the light beam either converges or diverges. Therefore, the half-angle of the detector should be larger than the divergence or convergence of the primary beam to ensure that all of the beam is received at the phototube.

3.6.4.   Electro-optical Phenomena in Colloidal Dispersions

Electrically induced optical phenomena can be very useful in the characterization of colloidal dispersions of anisotropic particles.

Anisotropy may not only result from the shape of the primary particles (cylinders, prolates or oblates, ellipsoids, etc.) but also when spherical particles coagulate into linear aggregates.

The electro-optical phenomena can be classified according to which optical effects are influenced by an electric field. These are electric light scattering, electric bire-fringence, and electric dichroism.

Electric light scattering is a combination of the orientation of anisotropic particles (or aggregates) in a pulsed electric field with light scattering. Under the influence of an electric field an anisotropy is induced in a dispersed system, which in turn induces anisotropy of the scattered light and therefore a dependence of the scattered light intensity on the angle of observation.

Electro-optic birefringence is concerned with the anisotropy of the refractive index of a colloidal dispersion under the influence of an applied electric field.

Electric dichroism is concerned with the anisotropy of the absorption coefficient of particles in an electric field. The optical density of a dispersion is different for light polarized perpendicular to that polarized parallel to the applied field.

The application of these methods to the investigation of coagulation phenomena is restricted to low electrolyte concentrations. This condition not only reduces Joule heating of the solution, but also assures greater electrical polarizability. This is because interfacial contribution is the largest and this decreases with increasing conductivity. Moreover, undesirable effects arising from electrode polarization are lessened. Another factor that must be considered is that linear coagulation may be promoted in an applied electric field [49, 50].

Most of the experimental arrangements for studying electro-optic phenomena have many features in common. In Figure 3.20 the apparatus of Oaklay and Jennings [51] for measuring electric birefringence is shown schematically. A He—Ne-laser L is used as the light source. The light beam passes through a high quality (Glan Thomson) prism P. The scattering cell C must be of good quality with strain-free windows. A pair of electrodes are placed in this cell. They are separated by a gap of about 2 mm. The electric field is established between them by means of a high potential pulse generator G. A pulsed voltage, giving rise to a low amplitude field, is applied to the cell. The electric field is horizontal across the sample and the initial polarizer P is adjusted to be at 45° polarization azimuth to the horizontal. In this way elliptical polarized light is obtained, which passes through a quarter wave plate Q and a second Glan-Thomson prism as an analyzer. The transmitted intensity is measured with a photo-multiplier PM. A transient recorder TR digitizes and stores the signals and a second channel records the pulse voltage applied to the sample. The data from the transient recorder can be output to a computer or oscilloscope or chart recorder.

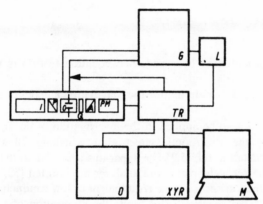

Figure 3.20
Scheme of the apparatus of Oaklay and Jennings [51].

The birefringence $\Delta v = v_{\parallel} - v_{\perp}$ is defined as the difference in the refractive index parallel and perpendicular to the optical axis of the polarized light.

The induced birefringence (Δv_0) of a colloidal dispersion is given by

$$\Delta v_0 = \frac{2\pi z}{v}(g_1 - g_2)\left[\frac{\mu^2}{k_B^2 T^2} + \frac{\Delta\alpha}{k_B T}\right]E^2 \tag{3.103}$$

v is the refractive index of the medium, z particle concentration, g_1, g_2 polarizability per unit volume of the particles parallel and perpendicular to the axis of symmetry, μ permanent dipole moment, $\Delta\alpha$ anisotropy in the electrical polarizability of the particles, E field strength.

For a polydisperse sample, or aggregated isodisperse sample of different aggregate sizes, the change in the birefringence (Δv_t) at any time t after the commencement of the decay (at $t = 0$ and $\Delta v_t = v_0$) is given by

$$\Delta v_t = \Sigma \, (\Delta v_0)_i \exp \left(-6 D_i^r t\right) \tag{3.104}$$

D^r is the rotary diffusion coefficient of aggregates of size i (or of particles of any other shape).

The rotary diffusion coefficient for particles of any shape is given by:

$$D_i^r = \frac{k_B T}{f} = \frac{k_B T}{\pi \eta L^3} \, \varphi(r) \tag{3.105}$$

with L the major particle dimension and r the axial ratio.

The function $\varphi(r)$ for a sphere is unity, for a prolate ellipsoid

$$\varphi(r) = \frac{3r^4}{2(r^4 - 1)} \left\{ \frac{2r^2 - 1}{r(r^2 - 1)^{1/2}} \ln \left[r + (r^2 - 1)^{1/2} \right] - 1 \right\} \tag{3.106}$$

and for a rod

$$\varphi(r) = 3(\ln 2r - 0.5) \tag{3.107}$$

If a linear aggregate is approximated by a rod then equation (3.101) is valid, with $r = a$.

The rotary diffusion plays a role in determining the coagulation kinetics of anisotropic particles.

Electric light scattering may also be applied in the investigation of linear aggregates by following the decay of the electrooptic effect after switching off the electric field. This was described in detail in ref. [52]. The apparatus is similar to that shown in Figure 3.20, if polarizer, quarter wave plate, and analyzer are omitted [52, 53].

One drawback of this method arises from the requirement of low conductivity for the dispersion to avoid electrode polarization. This limits the application of this method in coagulation studies, because it was shown [14] that slow coagulation may be reversible, which means that dilution of a coagulated dispersion may change the degree of coagulation.

It would be very interesting to carry out coagulation experiments using the same material, but where the particles differ in shape. This may be possible with certain types of metal oxide particles.

3.6.5.   Number Fluctuation Spectroscopy

The fluctuations in number and scattering power of particles in a small volume have been measured via the moments of the detected intensity distribution by Rarity and Randle [90]. Light from a helium—neon-laser is focused into the sample by a microscope objective. The image of the sample volume is transferred to a slit on the front end of a photoncounting photomultiplier tube via a microscope objective. The photomultiplier is mounted on a XYZ micropositioning stage, to allow alignment of the slit on the beamwaist. The scattering cell was thermostatable and mounted on a motor-driven micropositioning stage to allow translation of the sample perpendicular to the scattering plane at up to 4 mm · min^{-1}. The scattering volume was about $2 \cdot 10^{-8}$ mm^3. The puls train output from the photomultiplier electronics was fed through a ratemeter to a histogram analyzer. An internal microprocessor was programmed to calculate the normalized factorial moments.

3.7.   Experiments on Coagulation Kinetics

The development of the ultramicroscope by Siedentopf and Zsigmondy marked the first stage of investigations into coagulation kinetics, since this was really the first apparatus that enabled changes in the total number of particles with time to be measured. However, the coagulation still had to be "frozen" after chosen times. Continuous monitoring of the change in the number of particles with time became possible with the development of turbidity and light scattering methods, and then rapid particle counting (either electrical or optical) methods came about. These latter methods allow separate determinations to be made of the time evolution of singlets, doublets, triplets, and in some cases even higher aggregates.

3.7.1.   Experiments on Rapid Coagulation

The first experiments on rapid coagulation were performed by Zsigmondy [54], and by Westgren and Reitstötter [55], with colloidal gold particles. The gold dispersions were prepared by two methods, that is by the reduction of aurochloric acid (HAuCl$_4$) with either yellow phosphorus or with hydrogen peroxide. Westgren showed that only in the latter case "monodisperse" gold sols were obtained. To obtain dispersions of different particle size a different concentration of gold nuclei obtained with phosphorus was added to the stock solution of HAuCl$_4$. It was shown that the final number of gold particles was equal to the number of nuclei in this initial solution. In the coagulation experiments equal volumes of a gold sol and 0.2 mol · dm^{-3} sodium chloride solution were rapidly mixed. After various short time intervals the coagulation was stopped by adding 10 ml of 1% gum arabic solution or 10 ml of 10% gelatin solution. The particles were then counted using the ultramicroscope. In most of the experiments the initial particle concentration was between 10^8 and 10^{10} cm^{-3}.

To give some idea of the change in the total number of particles with time for different initial particle concentrations as predicted by the von Smoluchowski theory, some values are tabulated in Table 3.4. It is seen that, for particle concentrations of 10^9 cm^{-3} or greater, up to about 1 min there is virtually no change in the particle concentration. Considering that the lower limit of measuring time is in fact 1 min, the initial particle concentration should not greater than 10^9 parts · cm^{-3}.

Table 3.4

Change of the total particle number of initially monodisperse systems during coagulation according to Smoluchowski theory [equation (3.2)] at 298 K

t in s	n in Particles $\cdot$ cm^{-3}						
	10^6	10^7	10^8	10^9	10^{10}	10^{11}	10^{12}
0	10^6	10^7	10^8	10^9	10^{10}	10^{11}	10^{12}
30	$9.98 \cdot 10^5$	$9.97 \cdot 10^6$	$9.84 \cdot 10^7$	$8.62 \cdot 10^8$	$3.85 \cdot 10^9$	$5.90 \cdot 10^9$	$6.27 \cdot 10^9$
60	$9.98 \cdot 10^5$	$9.96 \cdot 10^6$	$9.68 \cdot 10^7$	$7.57 \cdot 10^8$	$2.38 \cdot 10^9$	$3.09 \cdot 10^9$	$3.13 \cdot 10^9$
180	$9.98 \cdot 10^5$	$9.89 \cdot 10^6$	$9.02 \cdot 10^7$	$4.80 \cdot 10^8$	$8.47 \cdot 10^8$	$9.17 \cdot 10^8$	$9.25 \cdot 10^8$
10^6	$1.57 \cdot 10^5$	$1.84 \cdot 10^5$	$1.87 \cdot 10^5$	$1.87 \cdot 10^5$	$1.87 \cdot 10^5$	$1.87 \cdot 10^5$	$1.88 \cdot 10^5$

From the results of Zsigmondy [54] and also those of Westgren and Reitstötter [55] and Tuorilla [56], the Smoluchowski rate constant k_S was calculated using equation (3.2),

$$k_S = \left(\frac{1}{z} - \frac{1}{z_0} \right) t^{-1} \equiv \frac{k_{S,\,theor}}{W} \qquad (3.108)$$

The theoretical value for the Smoluchowski constant in water solutions at 298 K equals

$$k_{S,\,theor} = 6.1 \cdot 10^{-12} \text{ cm}^3 \cdot \text{s}^{-1} \quad \text{at} \quad 298 \text{ K}$$

The experiments were performed at different particle concentrations from $2.69 \cdot 10^8$ to $2.02 \cdot 10^9$ particles $\cdot$ cm^{-3} [55] and from $6.3 \cdot 10^8$ to $3.5 \cdot 10^9$ particles $\cdot$ cm^{-3} [56]. The particle radii varied between 3.58 and 120 nm. The results are summarized in Figure 3.21. Because the original experiments were performed at different temperatures, the experimentally determined k_S values were converted to 298 K.

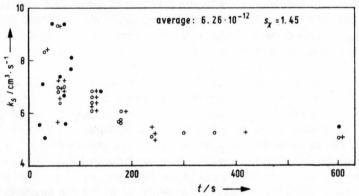

Figure 3.21
Calculated Smoluchowski rate constants k_S as a function of coagulation time of gold sols according to Zsigmondy [54], +; Westgren and Reitstötter [55], $\bigcirc$; and Tuorilla [56], $\bullet$. All the measurements were performed between 290 and 291 K, which means $k_{S,\,theor}$ = 5,85 $\cdot$ 10^{-12} cm^3 $\cdot$ s^{-1}.

From Figure 3.21 it can be seen that all the values for k_s scatter around a mean value of about $6.2 \cdot 10^{12}$ cm$^3 \cdot$ s^{-1}, with a standard deviation of 1.45. This mean value is very near to the theoretical value of $6.1 \cdot 10^{-12}$ cm$^3 \cdot$ s^{-1}.

Therefore one would postulate on this evidence that indeed the coagulation of gold particles does obey the Von Smoluchowski theory.

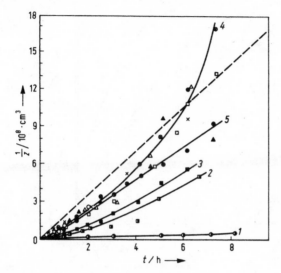

Figure 3.22
Dependence of the reciprocal concentration of gold particles on time of sodium chloride solution of different concentration according to Derjaguin and Kudravzeva [60].
1, $c = 1.8 \cdot 10^{-2}$ mol $\cdot$ dm^{-3}; 2, $c = 2 \cdot 10^{-2}$ mol $\cdot$ dm^{-3}; 3, $c = 3 \cdot 10^{-2}$ mol $\cdot$ dm^{-3}; 4, $c = 6 \cdot 10^{-2}$ mol $\cdot$ dm^{-3}; 5, $c = 10^{-1}$ mol $\cdot$ dm^{-3}. The dashed straight line corresponds to Smoluchowski theory.

The coagulation of gold particles was later investigated, also using the streaming ultramicroscope, by Derjaguin and Kudravzeva [59, 60]. Their red gold sol was prepared following the method of Zsigmondy. The particles had a mean radius between 20 and 30 nm, and the initial particle concentration was 10^{10} cm^{-3}. As coagulating electrolytes NaCl, MgSO$_4$, and La(NO$_3$)$_3$ were used. The results are shown in Figures 3.22 and 3.23. From the slope of the initial linear part of the coagulation curves one obtains the following values for the Smoluchowski rate constant: in NaCl solutions, $4.12 \cdot 10^{-12}$ cm$^3 \cdot$ s^{-1}, in MgSO$_4$ solutions $4.45 \cdot 10^{-12}$ cm$^3 \cdot$ s^{-1}, and in La(NO$_3$)$_3$ solutions $3.25 \cdot 10^{-12}$ cm$^3 \cdot$ s^{-1}. However, one should recall (Table 3.4) the time dependence of the particle concentration. For a particle concentration of 10^{10} cm^{-3} already after 3 min the particle concentration is reduced to $8.5 \cdot 10^8$ cm^{-3}. Because of this the actual value for the rate constant should be higher. Because the "incubation" time is rather high in experiments with the streaming ultramicroscope the initial Smoluchowski constant would be even higher than values (Fig. 3.21) obtained using the slit ultramicroscope. Nevertheless, given that it is coagulation in the primary minimum with gold particles of this size

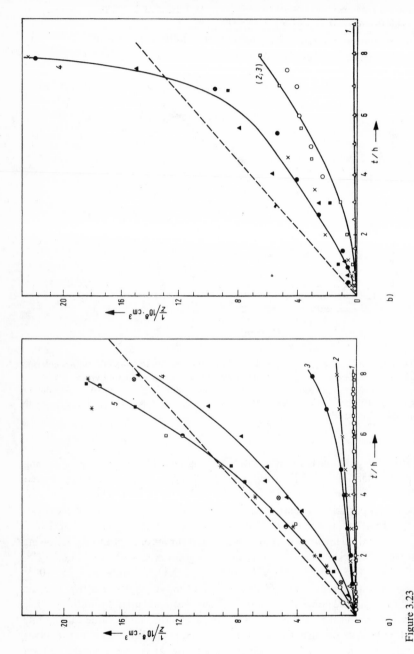

Figure 3.23

Reciprocal particle concentration plotted against time for gold sols flocculated with
(a) magnesium sulfate solution of different concentrations: 1, $c = 1.5 \cdot 10^{-4}$, $3.0 \cdot 10^{-4}$ and $3.5 \cdot 10^{-4}$ mol $\cdot$ dm^{-3}; 2, $c = 4 \cdot 10^{-4}$ mol $\times$ dm^{-3}; 3, $c = 5.0 \cdot 10^{-4}$ mol $\cdot$ dm^{-3}; 4, $c = 6 \cdot 10^{-4}$ mol $\cdot$ dm^{-3}; 5, $c = 8 \cdot 10^{-4}$, $10 \cdot 10^{-4}$, $1.5 \cdot 10^{-3}$, and $3 \cdot 10^{-3}$ mol $\cdot$ dm^{-3}.
(b) La(NO$_3$)$_3$: 1, $c = 1.5 \cdot 10^{-5}$ mol $\cdot$ dm^{-3}; 2 and 3, $c = 1.6$ and $2 \cdot 10^{-5}$ mol $\cdot$ dm^{-3}; 4, $c = 2.6$ and $4.0 \cdot 10^{-5}$ mol $\cdot$ dm^{-3} and $6.6 \cdot 10^{-5}$ mol $\cdot$ dm^{-3} according to Derjaguin and Kudravzeva [60].

this experimental coagulation rate constant is lower than the theoretically predicted value.

Using Zsigmondy's technique the coagulation of selenium sols was investigated by Kruyt and van Arkel [57, 58]. These sols were prepared by the reduction of $1 \text{ mol} \cdot \text{dm}^{-3}$ SeO_2 solution with a $1.5 \text{ mol} \cdot \text{dm}^{-3}$ hydrazine solution at 393 K. The initial particle concentration was $3.5 \cdot 10^{10}$ particles $\cdot \text{cm}^{-3}$ and the particle radii about 50—60 nm. The Smoluchowski rate constant was found to be in the range 1.2 to $2.1 \cdot 10^{-12} \text{ cm}^3 \cdot \text{s}^{-1}$. These values are much lower than the theoretical ones. The same system was later investigated by Watillon, Romerowski, and van Grunderbeek [61], the sol again being prepared by reduction with hydrazine. The initial particle concentration was between 10^8 and 10^9 cm^{-3}. The coagulation kinetics were investigated spectrophotometrically. In these experiments the Smoluchowski rate constant was found to be $2 \cdot 10^{-12} \text{ cm}^3 \cdot \text{s}^{-1}$, in very good agreement with Kruyt's value. Although the differences in the initial particle concentration in the two series of experiments were about two orders of magnitude, this is not significant, since the observation time was low in both cases.

Another, typically hydrophobic colloid, silver iodide, has been investigated by Ottewill [70]. The silver iodide sol was prepared by adding 125 cm^3 of a $1.25 \cdot 10^{-3} \text{ mol} \cdot \text{cm}^{-3}$ silver nitrate solution slowly to 100 cm^3 of $1.38 \cdot 10^{-3} \text{ mol} \cdot \text{dm}^{-3}$ potassium chloride solution, with vigorous stirring. The sol was electrodialyzed. The measurements were performed at pI 4. The sols were coagulated with $Ba(NO_3)_2$, $La(NO_3)_3$, and dodecyl pyridinium bromide (DPyBr). The Smoluchowski rate constant was found to be $3.2 \cdot 10^{-12} \text{ cm}^3 \cdot \text{s}^{-1}$ for $Ba(NO_3)_2$, $4.39 \cdot 10^{-12} \text{ cm}^3 \cdot \text{s}^{-1}$ for $La(NO_3)_3$, and $5.10 \cdot 10^{-12}$ $\text{cm}^3 \cdot \text{s}^{-1}$ for DPyBr. The particle radius was about 185 nm.

Ottewill [30] has also investigated rapid coagulation with arachidic acid sols of 100 to 250 nm radii with the streaming ultramicroscope. The Smoluchowski rate constant was found to be $2.7 \cdot 10^{-12} \text{ cm}^3 \cdot \text{s}^{-1}$.

Octadecanol dispersions with particle radii ~ 150 nm were prepared by the ultrasonic emulsification of molten octadecanol in pure water at 363 K. The relative coagulation rate constants were calculated from light scattering experiments and particle counting methods. The Smoluchowski rate constant was determined as $4.63 \cdot 10^{-12} \text{ cm}^3 \cdot \text{s}^{-1}$.

A typically hydrophilic colloid natural kaolin has been investigated by Tuorilla [56]. The larger particles were initially separated by centrifuging, so that the remaining sol was called "monodisperse". Some of the experimental data are collected in Figure 3.24. These were obtained by particle counting using the slit ultramicroscope after fixing the particles with gelatin. The average value of the Smoluchowski rate constant was $5.4 \cdot 10^{-12} \text{ cm}^3 \cdot \text{s}^{-1}$ at 288 K and $5.6 \cdot 10^{-12} \text{ cm}^3 \cdot \text{s}^{-1}$ at 298 K. These values are somewhat lower than the theoretically predcited ones. However, such a small difference between the theoretical and the experimental values is unexpected, since kaoline, being a hydrophilic colloid, should not coagulate to a deep primary minimum, since adsorbed water molecules on the particle surface will reduce the effective range of the van der Waals forces and will additionally produce repulsion forces, the so-called hydration forces. If we compare these values with those for gold particles also obtained using the slit ultramicroscope, it appears that the values obtained are dependent on the method of investigation.

7*

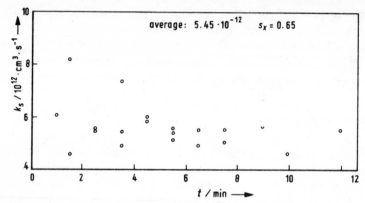

Figure 3.24
Calculated coagulation constants k_s from kinetic experiments with kaolin dispersions according to Tuorilla [56]

Studies on another hydrophilic system, quartz dispersions, have been described in ref. [62]. Since the particle size was lower than 1 μm the number of particle counts was made using the streaming ultramicroscope. The results are shown in Figure 3.25. From the initial slope of the $\frac{1}{z}$ versus t curves, for conditions of rapid coagulation a value for the Smoluchowski rate constant of $2.7 \cdot 10^{-11}$ cm$^3 \cdot$ s^{-1} is obtained, which is actually greater than the theoretical value.

In recent years many experiments have been carried out using polystyrene latices. These are good model dispersions because of their very good monodispersity and their

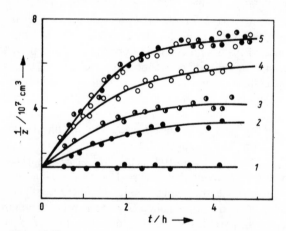

Figure 3.25
Plots of $1/z - t$ in quartz dispersions, particle diameter lower than 0.1 μm according to Černoberežskij et al. [62] in H$_2$O (1) and KCl solutions: 2, $c = 10^{-3}$ mol $\cdot$ dm^{-3}; 3, $c = 5 \cdot 10^{-3}$ mol $\cdot$ dm^{-3}; 4, $c = 10^{-2}$ mol $\cdot$ dm^{-3}; 5, $c = 4 \cdot 10^{-2}$, $8 \cdot 10^{-2}$, 10^{-1} mol $\times$ dm^{-3}.

ideal spherical shape. They are commercially available but may also be easily prepared in the laboratory.

If we look at the interaction energy—distance curve for latices (Figs. 1.15), then one can assume coagulation either in the primary or in the secondary minimum. A survey follows of the principal investigations with latices.

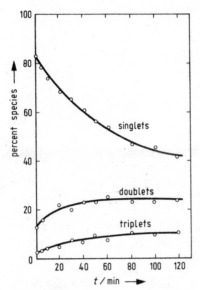

Figure 3.26
Time dependence of singlets, doublets, and triplets for polystyrene latex dispersion, $z_0 = 5.2 \cdot 10^7$ parts $\cdot$ cm^{-3}, $a = 915$ nm, flocculated with 1 % HCl according to Higuchi [39].

Higuchi et al. [39] investigated the coagulation of monodisperse "Dow" latices, with particle radii 915 nm, using the Coulter counter. They measured the time evolution of singlets, doublets, and triplets. Their results are shown in Figure 3.26. From these experiments the Smoluchowski rate constant for doublet and triplet formation can be calculated. For doublet formation $k_{s,11} = 3.5 \cdot 10^{-12}$ cm$^3 \cdot$ s^{-1}. From the time evolution of the doublets, that means taking into account triplet formation and decay of doublets; the Smoluchowski rate constant $k_{s,22}$ is $1.5 \cdot 10^{-12}$ cm$^3 \cdot$ s^{-1}. This value is lower than $k_{s,11}$. One would have expected just the opposite, considering the increase of the interaction energy in the secondary minimum. Higuchi et al. also determined the Smoluchowski rate constant for latices with different electrolytes. The results are shown in Figure 3.27. The k_s values appear to be independent of the nature of the coagulating electrolyte.

Matthews and Rhodes [40] investigated "Dow" latices with radii 357 and 600 nm, respectively. The kinetic experiments were again performed with the Coulter counter. The latices were coagulated with AlCl$_4$ solutions. Their results are shown in Figures

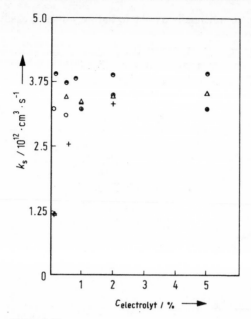

Figure 3.27
Experimental k_s values for polystyrene latex dispersions [39] with different electrolytes: ○, $MgSO_4$; ◓, $MgCl_2$; △, $NaCl$; +, Na_2SO_4.

3.28 and 3.29 for the two different latices. The following values for the Smoluchowski rate constants were found: $k_s = 4.07 \cdot 10^{-12}\ cm^3 \cdot s^{-1}$ for the 357 nm particles and $k_s = 2.68 \cdot 10^{-12}\ cm^3 \cdot s^{-1}$ for the 600-nm particles. This would seem to imply that the rate constant is dependent on the particle size. These authors also found the rate constant to be dependent on the particle concentration, at concentrations $< 5 \cdot 10^7\ cm^{-3}$. For example, k_s decreased to $2.55 \cdot 10^{-12}\ cm^3 \cdot s^{-1}$ at $3{,}5 \cdot 10^7\ cm^{-3}$ and to $1.4 \cdot 10^{12}\ cm^3 \cdot s^{-1}$ at $2 \cdot 10^7\ cm^{-3}$.

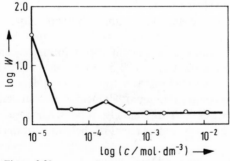

Figure 3.28
Stability ratio W as a function of aluminum chloride concentration for polystyrene latex dispersions $a = 357\ nm$ [40].

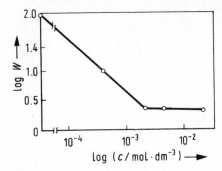

Figure 3.29
Stability ratio W as a function of aluminum chloride concentration for polystyrene latex dispersions $a = 600$ nm [40] ($\bigcirc$).

The same authors investigated the influence of surfactants on the coagulation rate constant. The results are shown in Figure 3.30. The Smoluchowski rate constant for rapid coagulation is not influenced by the presence of ammonium lauryl ether sulfate. Coagulation occurs here in the secondary minimum.

Hatton et al. [63] investigated four different polystyrene latices (with radii 175, 250, 430, 950 nm), prepared by emulsion polymerization following the method of Kotera et al. [64]. Their experimental method is described in detail in section 3.6.1. The coagulating electrolyte was $MgSO_4$. Their results suggest that coagulation kinetics are influenced by the initial particle concentration. However, in contrast to Matthews, they found a constant, limiting value at low particle concentrations. The results are summarized in Figure 3.31. With decreasing initial particle number concentration, the Smoluchowski rate constant approaches a limiting value of $2.6 \cdot 10^{-12}$ cm^3 · s^{-1}. No reason for this dependence was offered by the authors.

Lips and co-workers [68, 69] investigated the rapid coagulation of polystyrene latices using a low-angle light scattering technique. The rate constant, based on the change

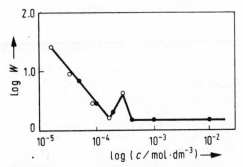

Figure 3.30
Stability ratio W as a function of aluminum chloride concentration for polystyrene latex dispersions $a = 357$ nm in the presence of $2 \cdot 10^{-5}$ mol · dm^{-3} ammonium lauryl sulfate [40]
($\bigcirc$, $\bullet$, separate experiments).

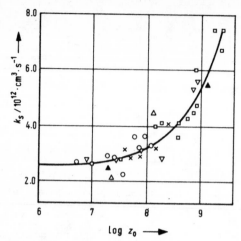

Figure 3.31

Smoluchowski rate constants k_s for rapid coagulation of polystyrene latex as a function of different initial particle concentrations z_0. $\square$, 185 nm radius; ∇, 250 nm radius; $\bigcirc$, 430 nm radius all 0.1 mol $\cdot$ dm^{-3} MgSO$_4$ at pH 5.5; $\triangle$, 950 nm radius, 0.1 mol $\cdot$ dm^{-3} MgSO$_4$ at pH 2.0; $\blacktriangle$, 950 nm radius, 0.2 molar KNO$_3$ at pH 5.5 according to Halton [63]; $\times$, 96 nm radius, 0.02 mol $\cdot$ dm^{-3} La(NO$_3$)$_3$ according to Rarity and Randle [90].

in the total number of particles, was calculated from the time dependence of the scattered light intensity, i.e., from the equation

$$\frac{i(t)}{I_0} = 1 + k_s z_0 t \tag{3.109}$$

Their results are summarized in Table 3.5. These results would suggest that k_S is independent of the particle size and the number of particle concentration.

Lichtenbelt et al. [44] followed the rapid coagulation kinetics of latices by monitoring the change in turbidity using a stopped flow spectrometer. The angle of illumination of the cuvette and the angle of acceptance of the photomultiplier tube were reduced to

Table 3.5

Smoluchowski rate constant for polystyrene latices as a function of particle radius and initial particle concentration

$a = 103$ nm		$a = 178$ nm		$a = 250$ nm	
z_0 in cm^{-3}	$k_S \cdot 10^{12}$ in cm$^3 \cdot$ s^{-1}	z_0 in cm^{-3}	$k_S \cdot 10^{12}$ in cm$^3 \cdot$ s^{-1}	z_0 in cm^{-3}	$k_S \cdot 10^{12}$ in cm$^3 \cdot$ s^{-1}
$2.2 \cdot 10^8$	3.4	$7 \cdot 10^6$	3.3	$3.53 \cdot 10^6$	3.4
$3.3 \cdot 10^8$	3.5	$1.05 \cdot 10^7$	3.2		
$4.4 \cdot 10^8$	3.4	$2.09 \cdot 10^7$	3.2		
		$3.14 \cdot 10^7$	3.3		

Table 3.6

Smoluchowski rate constant of polystyrene latices of different radii

a in nm	4.5	5.5	8.8	11.9	15.6	17.8
$k_s \cdot 10^{12}$ in $cm^3 \cdot s^{-1}$	3.4	3.2	3.6	2.8	2.1	2.9

$1°$. The mixing dead time was of the order of a few milliseconds. The lower limit of observation depended on the transmission change with time; a 1% transmission change could be measured with sufficient accuracy, in spite of the signal noise. The authors used polystyrene latices of different sizes both without added surfactants and with sodium dihexylsulfosuccinate present. No influence of the surfactant on the coagulation rate constant was observed. The results are summarized in Table 3.6.

The rate constant was found to be independent of the particle radius, as predicted from the theory of rapid coagulation, even when hydrodynamic interactions are taken into account. Van der Scheer et al. [72] determined the Smoluchowski rate constant of "Dow" latices using a stopped-flow technique. They found $k_s = 1.22 \cdot 10^{-12}$ $cm^3 \cdot s^{-1}$ at pH 8.9 and $1.15 \cdot 10^{-12}$ $cm^3 \cdot s^{-1}$ at pH 3.5 and $5 \cdot 10^{-2}$ $mol \cdot dm^{-3}$ $BaCl_2$. Gedan and co-workers [34, 66] have investigated the rapid coagulation kinetics of monodisperse polystyrene latices using the streaming ultramicroscope. The numbers of singlets, doublets, and triplets were monitored with time. The results are shown in Figures 3.32 and 3.33 for the condition $z_0 = 2 \cdot 10^8$ cm^{-3} and $c = 0.5$ $mol \cdot dm^{-3}$ KNO_3. The Smoluchowski rate constant for doublet formation was determined from the initial slope of the $z_1 - t$ curve as $1.3 \cdot 10^{-12}$ $cm^3 \cdot s^{-1}$. To obtain the rate constant for triplets and higher aggregate formation, the set of differential equations (3.3)—(3.5) was solved numerically. Using the values of Smoluchowski rate constants given in Table 3.7 the changes in the number of singlets (Fig. 3.34) and doublets, but not the number of triplets (Fig. 3.35) were described reasonably successfubly.

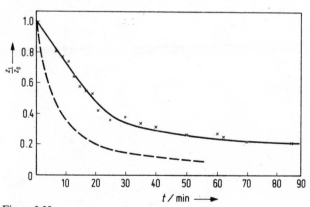

Figure 3.32

Decrease of the relative number of single particles z_1/z_0 with time [66]; dashed curve according to Smoluchowski theory.

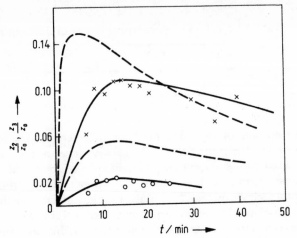

Figure 3.33
Change of the relative number of double z_2/z_0 (○) and triple z_3/z_0 (×) particles with time
[66]; dashed curve according to Smoluchowski theory.

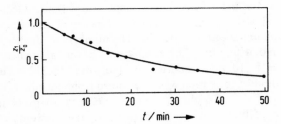

Figure 3.34
Decrease of the relative number of single particles with time (●) experimentally and calculated with the reaction constants in Table 3.7.

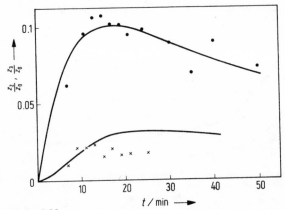

Figure 3.35
Change of the number of double (●) and triple (×) particles with time and calculated
with the reaction constants in Table 3.7.

The computations showed that only the values of k_{11} and k_{12} greatly influence the shape of the curves (Table 3.7). The higher rate constants could be varied between $8 \cdot 10^{-12}$ and $12 \cdot 10^{-12}$ cm$^3 \cdot$ s^{-1} without giving a better fit of the experimental data.

Table 3.7

Smoluchowski rate constants for different steps of coagulation [97]

Rate constants in cm$^3 \cdot$ s$^{-1} \cdot 10^{12}$	$k_{S, 11}$	$k_{S, 12}$	$k_{S, 13}$	$k_{S, 22}$	$k_{S, 23}$	$k_{S, 33}$
	2.4	3.7	8.7	12	12	12

Recently Cummins et al. [33] determined the rate constant for rapid coagulation of "Dow" latices using the streaming ultramicroscope. They obtained a much too high value of about 10^{-11} cm$^3 \cdot$ s^{-1}. Improved experiments [101] gave $k_S = 1.5$ cm$^3 \cdot$ s^{-1}.

From the about 40 different values for k_S of latices of different authors a mean value of $3.21 \cdot 10^{-12}$ cm$^3 \cdot$ s^{-1} is obtained with a standard deviation of $1.36 \cdot 10^{-12}$ cm$^3 \cdot$ s^{-1}.

To get a survey of the results of rapid coagulation, the value for k_S for different colloids are summarized in Table 3.8.

Table 3.8

Smoluchowski rate constants for different colloidal dispersions

Dispersion	Average value of $k_s \cdot 10^{12}$ cm$^3 \cdot$ s^{-1}	Standard deviation	Method
Gold	6.26	1.45	slit ultramicroscopy
Gold	3.9	—	particle counting
Selenium	2	—	slit ultramicroscopy and particle counting
Silver iodide	3.38	—	turbidity
Kaoline	5.45	0.65	slit ultramicroscopy
Arachidic acid	2.7	—	particle counting
Polystyrene	3.21	1.36	counting, turbidity, light scattering
Polyvinyl acetate	5.25	—	particle counting
Polyvinyl toluene	4.6	—	particle counting
Octadecanol	5.7	—	particle counting

From the experimental results on rapid coagulation one can draw the following conclusions:

— The experiments performed with the slit ultramicroscope give values for the Smoluchowski rate constant for gold particles somewhat higher than the theoretical Smoluchowski value, while for the (hydrophilic) kaoline particles the values obtained are slightly less than the theoretical value. Therefore, one can assume that this method gives too high a value for the coagulation rate constant.

— Particle counting methods and the stopped-flow technique for light scattering of turbidity measurements give lower values for k_S compared to the theoretical value, i.e., for gold particles about 62% less, for silver iodide particles about 54% less, and for polystyrene particles about 51% less.

If one calculates, from the experimental Smoluchowski rate constants, the coagulation probability, W, according to equation (3.108), i.e.,

$$W = \frac{k_{S,\text{theor.}}}{k_{S,\text{exp.}}}$$

together with equation (2.40)

$$N_0 = \frac{2}{W}$$

which is used for the estimation of the hydrodynamic interaction, one obtains the values given in Table 3.9.

Table 3.9

Coagulation probability W and N_0 for gold, silver iodide, and polystyrene latex dispersions, $N_{0,\text{calculated}}$ according to Honig et al. [75] including hydrodynamic interaction; see section 2.2.2.

Dispersed Phase	W	N_0	$N_{0,\text{calculated}}$
Gold	1.56	1.28	1.60
Silver iodide	1.80	1.11	1.60
Polystyrene	1.90	1.05	1.06

Assuming the hydrodynamic interaction to be the origin of the apparent retardation of rapid coagulation, one can calculate N_0 from the Hamaker constant (Table 2.1) or the B related values, i.e.,

$$B \equiv \frac{A}{12 k_B T}$$

where $B_{\text{polystyrene}} \sim 0.1$ and $B_{\text{Au, AgJ}} \sim 2$. The calculated N_0 values are tabulated in the third column of Table 3.9.

It would appear that at least for polystyrene latex hydrodynamic interactions could account for the observed retardation of coagulation.

However, for gold sols and silver iodide sols hydrodynamic interactions alone would not account for the degree of retardation observed. If this were the case, one would obtain from the coagulation experiments unreasonable values for the Hamaker constant, i.e., for gold $2.16 \cdot 10^{-20}$ J and for silver iodide $7 \cdot 10^{-21}$ J.

It might seem more reasonable therefore to explain the retardation of rapid coagulation in terms of the reversibility of coagulation as was first proposed by Martynov and Muller [13] and by Frens and Overbeek [73, 74].

However, another discrepancy then arises between theoretical predictions and experimental results—namely, reversibility implies that the rapid coagulation rate constant should be dependent on particle size [see equ. (3.57)]. But this has not been observed experimentally, with the exception of Matthews' results [40].

From the theoretical analysis of coagulation kinetics it follows that the rate constant should not depend on the initial particle number. However, Matthews and Rhodes [40] and Hutton et al. [63] showed experimentally with latices the rate constant to be dependent on the initial particle number. Hatton found (Fig. 3.31) that k_s increased from $2.6 \cdot 10^{-12}$ cm$^3 \cdot$ s^{-1} at $z_0 = 10^7$ cm^{-3} to $6.0 \cdot 10^{-12}$ cm$^3 \cdot$ s^{-1} at $z_0 = 10^9$ cm^{-3}. At $z_0 = 3 \cdot 10^9$ cm^{-3} the Smoluchowski rate constant was even higher than the theoretical predicted value.

Rarity and Randle [90] showed recently, using number-fluctuation spectroscopy, that the Smoluchowski rate constant increased from $2.8 \cdot 10^{-12}$ cm$^3 \cdot$ s^{-1} at $z_0 = 7 \cdot 10^7$ cm^{-3} to $k_s = 4.05 \cdot 10^{-12}$ cm$^3 \cdot$ s^{-1} at $6 \cdot 10^8$ cm^{-3}. Their results fit the curve of Hatton (see Fig. 3.31) very well.

Rapid coagulation of polystyrene latices containing the surfactant octa oxyethylene glycol n-dodecyl monoether in 10^{-1} mol $\cdot$ dm^{-3} MgSO$_4$ solution was investigated by Thompson [95] for different particle radii and at different temperatures using the flow ultramiscroscope. He obtained at 293 K for polystyrene particles, with radii 1.9 µm, $k_s = 6.7 \cdot 10^{-12}$ cm$^3 \cdot$ s^{-1} and with radii $a = 0,88$ µm, $k_s = 2.0 \cdot 10^{-12}$ cm$^3 \cdot$ s^{-1}. For the 1.9-µm particles at 313 K the Smoluchowski rate constant was lower than at 293 K, namely, $k_{s,313} = 2.76 \cdot 10^{-12}$ cm$^3 \cdot$ s^{-1}.

Vincent and co-workers [92—94] investigated rapid coagulation of polystyrene particles in the presence of adsorbed polyethylene oxide (PEO) by measuring the change in turbidity. They developed the following equation to relate the change in turbidity with time to the Smoluchowski rate constant and taking into account also the change in viscosity at different polymer concentration:

$$\left. \frac{d\left(\dfrac{\tau}{\tau_0}\right)}{d\left(\dfrac{l}{\eta}\right)} \right|_{t=0} = \frac{8k_B T z_0}{3W} \equiv \frac{2k_{s,0}}{W} = 2k_{s,\,eff} \tag{3.110}$$

$k_{s,0}$ is the measured Smoluchowski rate constant of the uncovered latex particles and $k_{s,\,eff}$ is the measured Smoluchowski rate constant in the presence of adsorbed macromolecules; η is the viscosity of the polymer solution.

The rate constant $k_{s,0}$ was found to be $2.9 \cdot 10^{-12}$ cm$^3 \cdot$ s^{-1}. The following results were obtained for PEO 20,000 in 0.5 mol $\cdot$ cm^{-3} NaNO$_3$ and at 298 K (Table 3.10). From these results it follows that with increasing polymer concentration rapid coagulation is strongly retarded. Above some critical value of the polymer concentration the dispersed system becomes stable. This critical polymer concentration depends strongly on the particle volume fraction. In the same paper the influence of temperature was investigated

Table 3.10

Smoluchowski rate constant $k_{S,\,eff} \cdot 10^{12}$ cm$^{-3} \cdot$ s of polystyrene latex with adsorbed PEO 20,000

Particle volume fraction	Conc. PEO in ppm					
	250	500	1000	1500	2000	2700
$1.5 \cdot 10^{-4}$	2.35	1.62	0.87	—	0.23	stable
10^{-4}	—	0.70	0.23	stable	—	—
$5 \cdot 10^{-5}$	—	0.58	stable	—	—	—

Table 3.11

Smoluchowski rate constant $k_{S,\,eff} \cdot 10^{12}$ cm$^{-3} \cdot$ s of polystyrene latex with adsorbed PEO 20,000

Temperature in K	Conc. PEO in ppm			
	500	1000	2000	3000
298	0.7	0.23	—	0.087
318	1.52	1.01	—	
338	—	1.09	0.47	0.29

on the Smoluchowski rate constant of polystyrene dispersions, stabilized with PEO 20,000. The results are tabulated in Table 3.11.

The effect of increasing temperature is to make the polymer dispersion less stable at a given polymer concentration. This is in agreement with an increase of the depth of the secondary (or primary) minimum by dehydration of the EO-chains.

Thompson [95] investigated rapid coagulation of polystyrene latex dispersions containing the surfactant octa oxyethylene glycol n-dodecyl monoether in 10^{-1} molar MgSO$_4$ solution at different particle radii and different temperatures. He obtained for particles of radii 1.9 µm in $1.8 \cdot 10^{-4}$ mol $\cdot$ dm^{-3} C$_{12}$(EO)$_8$ and 10^{-1} mol $\cdot$ dm^{-3} MgSO$_4$ at 298 K, $k_S = 6.7 \cdot 10^{-12}$ cm$^3 \cdot$ s^{-1}. For smaller particles ($a = 0,88$ µm) at 293 K the rate constant $k_S = 2.0 \cdot 10^{-12}$ was measured. At 313 K the rate constant for particles ($a = 0.88$ µm) was $2.76 \cdot 10^{-12}$ cm$^3 \cdot$ s^{-1}.

The increase of the Smoluchowski rate constant with increasing temperature can be explained by the dehydration of the EO-chains. This is connected with a increase of the minimum energy and therefore an decrease of the deaggregation probability.

3.7.2.   Experiments on Slow Coagulation

In slow coagulation not all the particle collisions are effective. There are several reasons for this. In the first place, with hydrophobic colloids such as metal sols or silver iodide sols, coagulation is essentially irreversible in the primary minimum, provided the particle size is small enough such that secondary minimum coagulation does not occur. Slow coagulation in this type of colloid in the primary minimum only occurs in the presence of an energy barrier.

With more hydrophilic particles such as oxide particles, slow coagulation because of an energy barrier may also occur. However, another factor leading to slow coagulation is reversibility. This may arise, for example, if adsorbed water molecules surround the particles, preventing direct contact of the particles. A third possible reason for slow, reversible coagulation is the presence of a secondary minimum and occurs typically with larger particles.

These different possibilities, which give rise to slow coagulation, are considered further below.

An energy barrier occurs when a sufficient electrostatic interaction is present. Only those collisions in which the mutual kinetic energy of the particles is greater than the height of the energy barrier will lead to coagulation.

In the DLVO theory an infinitely deep minimum is assumed for particles in direct contact. Martynow and Muller [13] and also Frens and Overbeek [73, 74] have shown that a relatively shallow minimum may result from the presence of adsorbed counterions or water molecules.

Deaggregation of doublets in the presence of an energy barrier in general has a low probability since the total energy barrier for this process is V_{max} plus V_{min}.

The typical hydrophobic colloids (Fig. 1.9) considered earlier in respect to primary minimum coagulation should, in general, remain irreversible even when an adsorbed monolayer of water or of specifically adsorbed ions is present, at least for particles whose radii exceed 20 nm. Smaller particles, however, may deaggregate if the distance of closest approach of 0.4 nm is introduced. Surface roughness or porosity may also reduce primary minimum energy.

Particles having a lower van der Waals attraction such as polymer latices (Fig. 1.15), have only a very small primary minimum, especially if one introduces the distance of closest approach. Therefore latex coagulation is mainly expected to occur in the secondary minimum, implying that the coagulation should be reversible.

Consideration is now given to slow barrier coagulation in the primary minimum in the presence of an energy barrier of limited height. As described in section 2.2.2., the coagulation probability, W, i.e., the ratio of the diffusion flux under slow and rapid coagulation conditions, depends on the total interaction energy, the van der Waals attraction and the hydrodynamic interaction, i.e.,

$$W = \frac{\displaystyle\int_0^\infty \frac{\beta(u)}{(u+2)^2} \left[\exp \left(\frac{V(u)}{k_B T} \right) \right] du}{\displaystyle\int_0^\infty \frac{\beta(u)}{(u+2)^2} \left[\exp \left(\frac{V_A(u)}{k_B T} \right) \right] du} \tag{2.41}$$

Numerical calculations of W have been performed by Honig et al. [75]. The influence of the van der Waals attraction is illustrated in Figure 2.1 a, in which the van der Waals energy is expressed as $B \equiv \dfrac{A}{12k_B T}$. The influence of the electrostatic repulsion on $\bar{W}$ is shown in Figure 2.1b. The electrostatic repulsion is expressed by G_{298} $= 11.21 \dfrac{a}{z^2} \tanh^2 (0.00973 z \psi_\delta)$ (with a in nanometers and ψ_δ in millivolts).

Recalling that W is defined as the ratio of the diffusion flux under slow and rapid coagulation conditions, one might expect that the hydrodynamic interaction is eliminated since one would expect that the hydrodynamic correction under rapid and slow coagulation is nearly the same.

Hence it follows that the slow coagulation should be proportional to the particle radius and inversely proportional to the square of the valency of the counterions. However, some exceptions for these predictions were observed by Reerink and Overbeck [76] (see Fig. 2.2). If the energy maximum is much greater than $k_B T$, only those values of $\dfrac{V}{k_B T} (u + 2)^{-2}$ that are in the neighborhood of the maximum, at the dimensionless distance $u_m \left(u_m = \dfrac{d_m}{a} \right)$, contribute significantly to the Fuchs integral. Therefore, one may write

$$W = \frac{2}{(u_m + 2)^2} \int_0^\infty \exp \left(\frac{V}{k_B T} \right) du \tag{3.111}$$

Neglecting possible contributions from hydrodynamic interaction Reerink and Overbeck [76] obtained the following relationship between log W and log c:

$$\log W = - \frac{A}{12 u_m k_B T} \log c - \frac{A}{24 u_m k_B T} \log \left[8\pi N_L z^2 e^2 a^2 10^{-6} (\varepsilon k_B T)^{-1} \right]$$

$$+ \left(\frac{3}{2} - \frac{A}{12 u_m k_B T} \right) \log u_m + \frac{1}{2} \log \left(\frac{96 k_B T \pi}{A} \right)$$

$$- \frac{1}{2} \log (2 - \varkappa a u_m) (u_m + 2)^4 \tag{3.112}$$

Since, with varying electrolyte concentration, the changes in $(2 - \varkappa a u_m)$ und u_m are small (see Figs. 1.9, 1.12, and 1.15) all the terms in equation (3.112) may be considered to be independent of the electrolyte concentration with the exception of the first term. Therefore, one may write

$$\log W = -K_1 \log c + K_2 \tag{3.113}$$

Honig et al. [75] included the hydrodynamic interaction; the following term must then be added to equation (3.112):

$$\frac{(1 - \varkappa d_m) (4 + 12 \varkappa d_m) + 27 (\varkappa d_m)^2}{2(2 - \varkappa d_m) (2 + \varkappa d_m) (1 + 6 \varkappa d_m) (2 + 3 \varkappa d_m)}$$

According to equation (3.112) the slope of a log W/log c plot is given by the following equation:

$$\frac{d \log W}{d \log c} = \frac{1}{2} \frac{d \log W}{d \log a} = -2.15 \cdot 10^7 \frac{a}{z^2} \tan h^2(0.009\,73\; z\psi_\delta) \qquad (3.114)$$

a is expressed in cm and ψ_δ in millivolts, z is the valency of the counterions.

Table 3.12

Values of the slope of log W/log c curves for different dispersed systems, as a function of particle radius and counterion valence

Sol	d log W/d log c	Particle radii in nm	Electrolyte	Reference
Silveriodide	−10.6	12.5	KNO_3	[76]
	− 5.9	26.0	KNO_3	[76]
	− 7.3	100.0	KNO_3	[76]
	− 8.0	12.5	$Ba(NO_3)_2$	[76]
	− 8.0	26.0	$Ba(NO_3)_2$	[76]
	− 8.0	32.5	$Ba(NO_3)_2$	[76]
	−11.0	100.0	$Ba(NO_3)_2$	[76]
	−15.8	26.0	$La(NO_3)_3$	[76]
Selenium	−13.0	25.0	KCl	[57, 58]
	− 4.7	25.0	KCl	[57, 58]
	− 5.8	31.0	KCl	[57, 58]
	− 5.0	25.0	$BaCl_2$	[57, 58]
	− 6.2	25.0	$BaCl_2$	[57, 58]
	−11.8	31.0	$BaCl_2$	[57, 58]
	− 2.3	70.5	KCl	[61]
	− 2.1	84.0	KCl	[61]
	− 2.1	94.0	KCl	[61]
	→ 2.0	102.0	KCl	[61]
	− 1.7	91	KCl	[61]
	− 2.4	91	$BaCl_2$	[61]
	− 1.4	91	$LaCl_3$	[61]
Gold	−14.0	35.0	$LiCl$	[56]
	− 4.0	90.0	$LiCl$	[56]
Latex	− 2.42	300	$Ba(NO_3)_2$	[41]
	− 2.78	515	$Ba(NO_3)_2$	[41]
	− 2.71	1213	$Ba(NO_3)_2$	[41]
	− 2.70	1840	$Ba(NO_3)_2$	[41]
	− 1.59	2115	$Ba(NO_3)_2$	[40]
	− 0.9	120	$Ba(NO_3)_2$	[40]
	− 0.34	714	$Ba(NO_3)_2$	[40]
	− 3.11	30	HNO_3	[65]
	− 1.35	51	HNO_3	[65]
	− 1.15	121	HNO_3	[65]
	− 0.98	435	HNO_3	[65]

From equation (3.114) it follows that the slope of the $\log W/\log c$ plot should be proportional to a, and vary with the valency of the counterions as z^{-2}, i.e., in the ratio

$$1:0.25:0.11$$

Some results from different authors are summarized in Table 3.12.

The data, in general, do not support the Reerink-Overbeck predictions. There may be several reasons for this. For example, as mentioned above, latices are not expected to coagulate in the primary minimum and so reversibility cannot be neglected.

One could extend the theory to incorporate deaggregation, then W has to be multiplied by the factor

$$\left(\frac{1}{1 - \exp \dfrac{V_{min}}{k_B T}} \right)$$

according to equation (2.48). If we express the retardation of coagulation by the ratio of the aggregation probability α and deaggregation probability β, then

$$\frac{\alpha}{\beta} = 16\pi a^2 \delta_2 \left[\exp\left(-\frac{|V_{min}|}{k_B T} \right) - 1 \right] \tag{3.57}$$

Hence the retardation factor should depend on a^2.

Reversibility of coagulation by changing the composition of the dispersion medium has been observed for various kinds of colloidal dispersions such as silver iodide sols [74, 77, 78], oxide sols [62, 79], and polymer latices [14, 80, 81] dispersions.

The work of Benitez and Ritchie [79] may be used to illustrate the effects of reversibility. They showed that the flocculation of ferrous oxide (Fig. 3.36a) and Prussian blue particles (Fig. 3.36b) was fully reversible when the flocculated sols were diluted with the electrolyte-free, unflocculated sols. Vanadium pentoxide (Fig. 3.36c) and arsenic sulfide sols did not coagulate fully reversibly. Vincent et al. [80] investigated the slow reversible coagulation of latices stabilized with n-dodecyl hexaoxyethylene monoether. The experiments were carried out wich polystyrene latices of different particle size. Barium chloride was added as coagulant, at a concentration $0.1 \text{ mol} \cdot \text{dm}^{-3}$ was chosen to be in excess of the flocculation concentration. After 48 h equilibrium was established. Slow rotating prevented sedimentation. The optical density OD was measured as a function of the particle volume fraction. The results are shown in Figure 3.37 for five latices of different size. In these experiments $\dfrac{d \log OD}{d \log \lambda}$ was plotted as a function of particle concentration (λ is the wavelength of light, which was varied from 400 to 600 nm). This reaction is a sensitive function of particle size. The breaks in the curve indicated coagulation. The flocs were in true equilibrium with single particles. After dilution the flocs spontaneously broke up into singlets.

Latex 1 with the smallest particles did not coagulate at any concentration. With the other latices a crit.cal volume fraction was found, above which reversible coagulation was obtained. By analogy with the vapor-liquid phase transition the Boltzmann relation-

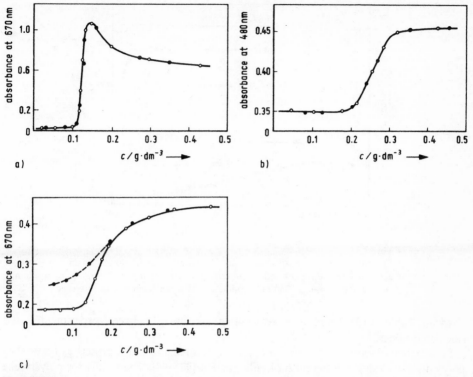

Figure 3.36
Flocculation of different sols: (a) hydrous ferric oxide sol by sodium sulfate; (b) Prussian blue sol by barium chloride; (c) vanadium pentoxide sol by barium chloride, ○, increasing, ●, decreasing, electrolyte concentration according to reference [79].

ship was used to relate the volume concentration of particles in the dispersed phase φ_D to that in the floc φ_F phase at equilibrium, i.e.,

$$\varphi_D = \varphi_F \exp\left(-\frac{V_{min}}{k_B T}\right) \tag{3.115}$$

if one assumed only linear contacts are formed. In compact aggregates more than one "bond" has to be broken and in that case the term in brackets has to be multiplied by the number of bonds, n. φ_F should be somewhere in the range of 0.3—0.7, depending on the rates of aggregation and deaggregation of the flocs. If one assumes that, at the break points in the curves (Fig. 3.37), only doublets are formed, then

$$\varphi_{crit} = \varphi_F$$

From the experiments the depth of the energy minimum was calculated. These values were compared with theoretical values for the van der Waals attraction assuming a distance of closest approach of either 0.3 or 1 nm. The van der Waals energy was calculated using the expression of Vold and co-workers [82, 83]. The results are summa-

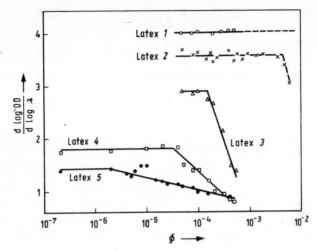

Figure 3.37
Logarithm optical density divided by log wavelength is plotted against the volume fraction of five different latices of different size: *1*, $a = 22$ nm; *2*, $a = 60$ nm; *3*, $a = 110$ nm; *4*, $a = 250$ nm; and *5*, $a = 490$ nm according to reference [80].

rized in Table 3.13. There is a reasonable agreement between experimental and calculated values.

Slow coagulation of polystyrene latices was measured also in the presence of adsorbed poly(ethylene oxide) of different molecular weight [92—94]. The results for polystyrene particles of radii of 95.5 nm and PEO 1500 are shown in Figure 3.38. The initial increase of coagulation rate, below saturation concentration, is due to bridging flocculation, i.e., flocculation resulting from the coadsorption of polymer onto more than one particle. The decrease in W^{-1} beyond the maximum is due presumably to the buildup in polymer adsorption from the point of monolayer coverage to the plateau level of adsorption. From the plateau values of W^{-1} the energy of the minimum was calculated with equation (2.40). The results are tabulated in Table 3.14.

Table 3.13

Comparison of V_{min} values from weak coagulation experiments and from theoretical calculations according to Vincent and co-workers [80]

Particle radii in nm	$\dfrac{V_{min}}{k_B T}$		$\dfrac{V_{min}}{k_B T}$ linear aggregates
	$\delta = 1$ nm	$\delta = 0.3$ nm	
60	1.5	3.8	4.2
110	2.6	6.8	7.5
250	6.4	15.8	9.2
490	12.3	30.8	11.7

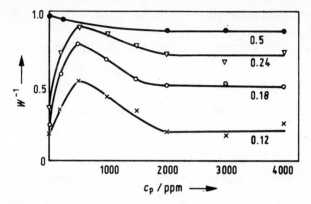

Figure 3.38

W^{-1} versus PEO 1500 concentration c_p at various sodium nitrate concentration (in mol $\cdot$ dm^{-3}) as indicated: 298 K and volume fraction of the particles $\varphi = 10^{-4}$.

Table 3.14

Limiting W^{-1} and calculated $\dfrac{V_{min}}{k_B T}$ for latex plus PEO 1500

c_{NaNO_3} in mol $\cdot$ dm^{-3}	W^{-1}	$\dfrac{V_{min}}{k_B T}$
0.12	0.21	—0.24
0.18	0.52	—0.75
0.24	0.73	—1.31
0.50	0.88	—2.12

Smith and Thompson [81] also investigated aggregation equilibrium in polystyrene latices and in graphitized carbon dispersions using streaming ultramicroscopy. The aim of this work was also to relate the equilibrium state to the pairwise van der Waals attraction for latex particles. They considered the process

$$n \text{ mer} + j \text{ mer} \underset{k_{-1}}{\overset{k_1}{\rightleftharpoons}} (n + j) \text{ mer}$$

At equilibrium

$$k = \frac{[(n + j) \text{ mer}]}{[n \text{ mer}] [j \text{ mer}]} \tag{3.116}$$

If k_1 and k_{-1} are the respective rates of the forward and reverse processes, then

$$k = \frac{k_1}{k_{-1}} \tag{3.117}$$

They also considered linear aggregation with the simplification that an i mer has $n = (i - 1) z_i$ bonds.

By defining k specifically in terms of the singlet/doublet process, we obtain

$$k = \frac{z_2}{z_1^2} \tag{3.118}$$

If one introduces instead of the number of particles the number of bonds then

$$k = \frac{n_2}{z_1^2} = \frac{n_2}{(z_0 - 2n_2)^2} \tag{3.119}$$

where $z_1 = z_0 - 2n_2$. If one compares equation (3.119) with equation (3.54) it follows that

$$k \approx \frac{1}{2} \frac{\alpha}{\beta} \tag{3.120}$$

Smith et al. have found equilibrium coagulation; a typical plot of reciprocal particle concentration against time for 0.15 μm radii graphitized carbon particles in 10^{-2} mol $\cdot$ dm^{-3} KNO$_3$ is shown in Figure 3.39. From the initial slope of this curve an energy barrier of about 9 kT was calculated. This shows that coagulation was in the primary minimum.

The variation of equilibrium particle concentration was investigated for polystyrene latex of radii 0.8 μm in 10^{-2} mol $\cdot$ dm^{-3} KNO$_3$ and for graphitized carbon of radii 0.15 μm in 10^{-2} mol $\cdot$ dm^{-3} KNO$_3$ as a function of the initial particle concentration (Fig. 3.40). The results are tabulated in the Tables 3.15 and 3.16.

The minimum energy is far from independent of particle concentration, the variation being such that with increasing particle concentration the minimum energy becomes lower. This is the opposite effect from that which would result from nonlinear aggregates.

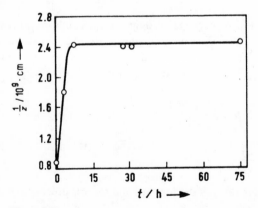

Figure 3.39
Reciprocal particle concentration against time for 150-nm radii particles of graphitized carbon in 10^{-2} mol $\cdot$ dm^{-3} KNO$_3$ according to reference [81].

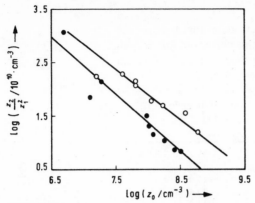

Figure 3.40
Variation of the doublet/singlet ratio z_2/z_1^2 with equilibrium particle concentration. $\bigcirc$, polystyrene latex of radius 800 nm in 10^{-2} mol $\cdot$ dm^{-3} KNO$_3$; $\bullet$, 150-nm graphitized carbon in 10^{-2} mol $\cdot$ dm^{-3} KNO$_3$ according to ref. [81].

Table 3.15

Minimum energy as a function of the initial particle concentration for polystyrene latex of radii 0.8 μm in 10^{-2} mol $\cdot$ dm^{-3} KNO$_3$

Initial particle concentration in 10^7 cm^{-3}	Minimum energy $k_B T$ units
1.6	−11.8
4.4	−11.9
6.0	−11.6
6.0	−11.4
11.5	−10.7
18.8	−10.6
38.0	−10.3
60.0	− 9.4

In the same paper also mixtures of two polystyrene latices of radii 0,8 μm (weakly aggregating) and 0.04 μm (stable) and 0.125 μm (stable) in 10^{-2} mol $\cdot$ dm^{-3} KNO$_3$ were described (Fig. 3.41). From their data the minimum energy was calculated (Tables 3.17; 3.18).

The position of equilibrium coagulation was found to depend on the concentration of smaller particles in both cases.

In the same paper also the influence of the temperature on the coagulation rate constant was measured. The increase in temperature brings about an increase in aggregation at all particle concentrations, i.e., the depth of the energy minimum increases.

Table 3.16

Minimum energy as a function of the initial particle concentration for graphitized carbon particles of radii 0.15 μm in 10^{-2} mol · dm^{-3} KNO$_3$

Initial particle concentration in 10^7 cm^{-3}	Minimum energy $k_B T$ units
0.5	−17.2
1.25	−13.7
2.0	−15.0
9.5	−13.4
10.0	−13.0
12.5	−12.6
18.2	−12.3
26.0	−11.9
32.0	−11.9

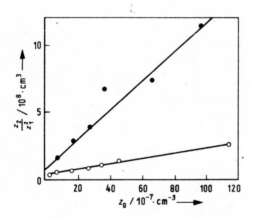

Figure 3.41
Plot of variation of the doublet/singlet ratio z_2/z_1^2 against z_0 ($= 1.02 \cdot 10^8$ cm^{-3}), polystyrene latex of radii 0.8 μm in 10^{-2} mol · dm^{-3} KNO$_3$ with added ○ 0.04 μm polystyrene latex dispersions and ● 0.125 μm polystyrene latex according to ref. [81].

Table 3.17

Minimum energy of 0,8-μm polystyrene latex ($z_0 = 1.02 \cdot 10^8$ cm^{-3}) in 10^{-2} mol · dm^{-3} KNO$_3$ with added 0.125-μm polystyrene latex at 298 K

z_t in 10^7 cm^{-3}	7.2	16.9	28.2	37.0	67.0	96.0
$\dfrac{V_{min}}{k_B T}$ units	−11.7	−12.3	−12.6	−13.1	−13.2	−13.6

Table 3.18

Minimum energy of 0.8-μm polystyrene latex ($z_0 = 1.02 \cdot 10^8$ cm^{-3}) in 10^{-2} mol · dm^{-3} KNO$_3$ with added 0.04-μm polystyrene latex at 298 K

z_t in 10^7 cm^{-3}	3	8	16.9	27.4	35.5	46.8	115
$\dfrac{V_{min}}{k_B T}$ units	−10.5	−10.7	−10.9	−11.1	−11.4	−11.6	−12.1

This may be consistent with a primary minimum restricted by a simple stand-off of ca. 1 nm. Conversely the data do not appear consistent with secondary minimum coagulation.

Thompson and Pryde [96] also investigated weak aggregation of polystyrene dispersions, stabilized with *n*-dodecyl octa ethylene glycol monoether. The electrolyte concentration was high enough to suppress electrostatic interaction. They have measured the influence of the initial particle number concentration (Fig. 3.42) and of the temperature (Fig. 3.43) on the coagulation rate. They have found that independent of the initial particle concentration and of the temperature equilibrium was established. From their data $\dfrac{\alpha}{\beta}$ and the minimum energy were calculated (Table 3.19).

It follows that in these experiments the energy minimum was dependent on the initial particle number.

In the same paper the effect of temperature of nonionic surfactant stabilized dispersions was measured for different particle sizes. The results are tabulated in Table 3.20, Table 3.21, and Table 3.22.

It follows, as expected, that with increasing temperature the minimum energy becomes greater.

Thompson [95] also investigated slow coagulation of polystyrene latex ($a = 0.5$ μm) in the presence of $1.86 \cdot 10^{-1}$ mol · dm^{-3} C$_{12}$EO$_8$ in 10^{-4} molar MgSO$_4$ at 308 K. The

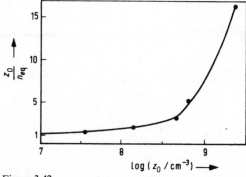

Figure 3.42

Plot of degree of aggregation against particle concentration for polystyrene latex dispersions of radii 0.5 μm in 10^{-1} mol · dm^{-3} MgSO$_4$ stabilized by a monolayer of C$_{12}$EO8 according to reference [81].

result is shown in Figure 3.44. From these data an energy minimum $-9.4k_BT$ and an energy barrier of $9.7k_BT$ were calculated.

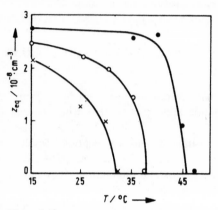

Figure 3.43

Plot of equilibrium aggregation against temperature for $C_{12}EO8$ stabilized latex dispersions in 10^{-1} mol $\cdot$ dm^{-3} MgSO$_4$: ●, 0.125 μm; ○, 0.25 μm; and ×, 0.8 μm radius.

Table 3.19

Minimum energy of aggregated polystyrene particles of radii 0.5 μm in 10^{-1} mol $\cdot$ dm^{-3} MgSO$_4$, stabilized by a monolayer of $C_{12}EO_8$ assuming linear aggregation [96]

$z_0 \cdot 10^{-8}$ cm^{-3}	25.1	6.9	3.3	1.45	0.30
$\dfrac{V_{min}}{k_BT}$	-25.1	-22.7	-21.5	-20.3	-18.4

Table 3.20

Minimum energy of polystyrene latex with $a = 0.175$ μm and $z_0 = 3 \cdot 10^8$ cm^{-3}

T in K	$n_{eq} \cdot 10^{-8}$ cm^{-3}	$\dfrac{V_{min}}{k_BT}$
288	2.7	-15.3
308	2.5	-15.5
313	2.6	-15.4
318	0.85	-15.8

Table 3.21

Minimum energy of polystyrene latex with
$a = 0.5\ \mu m$ and $z_0 = 3 \cdot 10^8\ cm^{-3}$

T in K	$n_{eq} \cdot 10^{-8}\ cm^{-3}$	$\dfrac{V_{min}}{k_B T}$
288	2.5	−13.4
308	2.2	−13.7
313	2.0	−14.0
318	1.4	−14.5

Table 3.22

Minimum energy of polystyrene latex with
$a = 0.8\ \mu m$ and $z_0 = 3 \cdot 10^8\ cm^{-3}$

T in K	$n_{eq} \cdot 10^{-8}\ cm^{-3}$	$\dfrac{V_{min}}{k_B T}$
288	2.13	−13.0
308	1.25	−14.2
313	0.97	−14.7

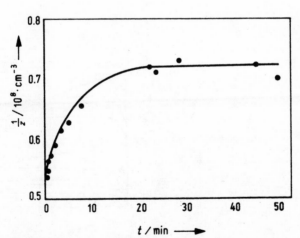

Figure 3.44

Aggregation of polystyrene latex (radius 0.5 μm) dispersions in $1.86\ mol \cdot dm^{-3}\ C_{12}EO8$ and $10^{-4}\ mol \cdot dm^{-3}\ MgSO_4$. Theoretical curve based on the initial Smoluchowski rate constant $k_S = 7 \cdot 10^{-14}$ and equilibrium ratio $z_2/z_1^2 = 2.31 \cdot 10^{-9}\ cm^3$.

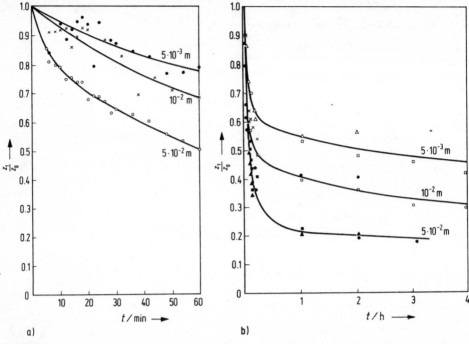

a)   b)

Figure 3.45
Plots of the variation of the dimensionless number of singlet particles (polystyrene of radii $a = 0.34 \cdot 10^{-4}$ cm) with time at different electrolyte concentrations and at the different time scale according to reference [97].

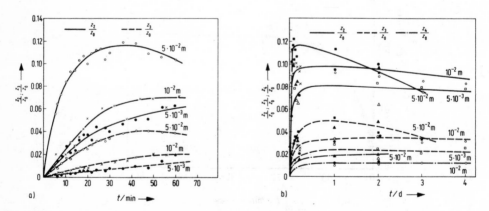

a)   b)

Figure 3.46
Plots of variation of the dimensionless number of doublets, triplets, and quadruplets (polystyrene of radii $a = 0.34 \cdot 10^{-4}$ cm) at different electrolyte concentrations and at different time scale according to reference [97].

Table 3.23

α, $\dfrac{\alpha}{\beta}$, V_{max} and V_{min} for slow coagulation of polystyrene latex. Particle radius $0.30 \cdot 10^{-4}$ cm at three different electrolyte concentrations

c_{el} in mol $\cdot$ dm^{-3}	$\dfrac{\alpha}{\beta}$ in cm^3	$\dfrac{V_{min}}{k_B T}$	α in cm$^3 \cdot$ s^{-1}	$\dfrac{V_{max}}{k_B T}$
$5 \cdot 10^{-3}$	$(0.7 \ \pm 0.12)\,10^{-9}$	$-\ 9.1$	$(4.8 \pm 0.5)\,10^{-13}$	6.6
10^{-2}	$(0.97 \pm 0.22)\,10^{-9}$	-10.1	$(8.5 \pm 0.7)\,10^{-13}$	6.8
$5 \cdot 10^{-2}$	$(2.8 \ \pm 0.6)\,10^{-9}$	-11.3	$(17 \pm 2.7)\,10^{-13}$	6.8

Recently Sonntag et al. [97] investigated slow coagulation of polystyrene latices at different electrolyte concentrations using streaming ultramiscroscopy. Plots of the dependence of the number of singlets against time are shown in Figure 3.45 for three different electrolyte concentrations in different time scales. Plots of the dependence of the number of bonds for doublets, triplets, and quadruplets against time (also in different time scales) at three different electrolyte concentrations are shown in Figure 3.46. From Figure 3.46 it follows that no real equilibrium was obtained. Therefore the curves were analyzed as follows. If one takes from the curves of the number of bonds versus time at two different times the numbers of bonds and the slopes dn_1/dt, dn_2/dt, and so on then one obtains α and β values and hence $\dfrac{\alpha}{\beta}$ too. The average values for both quantities are shown together with the values for the energy barrier and energy minimum, respectively, in Table 3.23.

It follows that as expected the collision efficiency α increases with increasing electrolyte concentration. However, the energy barrier remains constant or even increases. It follows that the use of the Debye length as an approximation for the width of the energy barrier is incorrect. That is seen also from the energy distance curve in Figures 1.15 and 1.24. May be that structural forces are responsible for a minimum of restricted depth.

On the other hand, for the calculation of the minimum energy the width of the minimum can be approximated by the Debye length.

3.7.3.   Experiments on the Kinetics of Bridging Flocculation

Polymeric flocculants are frequently used in solid-liquid separations. It is generally agreed that these molecules act by adsorption of a single molecule at the surface of two particles, i.e., by the formation of polymer bridges between them. The rapid coagulation rate for partially macromolecule coated particles should therefore be larger than the rapid coagulation rate for bare particles. The enhanced rate of aggregation, i.e., the higher collision frequency, can be explained by a reduced hydrodynamic interaction between the partially covered particles and by their increased collision radius [$R > 2a$ in equ. (2.29)]. According to equation (3.21 a) the coagulation rate depends on the product of collision frequency and collision effectivity. According to Latter and co-workers [98, 99] the efficiency factor α_{ef} could be estimated from the following equation:

$$\alpha_{ef} = \Theta(1 - \Theta) \tag{3.121}$$

where Θ is the fractional coverage of the solid surfaces by adsorbed polymer. In effect Θ represents the probability that, in the region of contact, the surface of one particle is covered by polymer while $(1 - \Theta)$ describes the probability that the surface of one particle is bare. Equation (3.121) predicts that the maximum collision efficiency will occur, when $\Theta = \dfrac{1}{2}$ and will have a value of $\dfrac{1}{4}$. It was recently shown by Hogg [100] that several criticisms can be leveled at the efficiency factor, defined by equation (3.121). Optimum flocculation appears to occur at fractional coverages significantly less than one-half. In this paper the collision efficiency was expressed by

$$\alpha_{ef} = 1 - \Theta^{n_i + n_j} - (1 - \Theta)^{n_i + n_j} \tag{3.122}$$

if the adsorbed molecules are distributed at random over the n_i (n_j, respectively) sites on the surface of particle i (j, respectively). According to equation (3.108) the stability ratio is defined as

$$W = \frac{k_{S,\,theor}}{k_{S,\,exp}} \equiv \frac{\alpha_{ef,\,theor}}{\alpha_{ef,\,exp}}$$

Under the condition of bridging flocculation W should become lower than 1.

Van der Scheer et al. [72] investigated bridging flocculation of polystyrene latex with human serum albumin (HSA) and fibrinogen (HFb). They investigated the influence of the HSA and HFb concentration, the influence of the electrolyte concentration, and the influence of the pH on the coagulation rate. Under optimum coagulation conditions the maximum flocculation rate of polystyrene particles with adsorbed HSA at $5 \cdot 10^{-2}$ mol $\cdot$ dm^{-3} BaCl$_2$ was $k_S = 1.62 \cdot 10^{-12}$ cm$^3 \cdot$ s^{-1} compared to the maximum coagulation of the bare particles $k_S = 1.22 \cdot 10^{-12}$ cm$^3 \cdot$ s^{-1}. There is an acceleration by 1.3 times. The increase in the rate constant by addition of HFb equals 1.6 times the rate constant of the bare particles. However, the rate constant under bridging flocculation is only about 25% of the theoretical Smoluchowski rate constant.

4.   Structure Formation in Disperse Systems

Up to this point only the preliminary steps of coagulation in very diluted systems have been considered. In such dispersions, since the number of particles is very small ($10^8 - 10^{10}$ cm^{-3}), the coaguli can only grow to a limited extent. If they are able to grow to large sizes, they may then settle, forming a sediment. Sediments created in this manner are examples of a "structured" disperse system, which is the topic to be discussed in this chapter.

A more interesting type of a structured disperse system occurs when the initial particle concentration is high enough that a continuous structure forms through the whole bulk, either after the onset of coagulation or by virtue of mutual repulsion forces between the particles. As will be shown later, the necessary minimum particle concentration depends on the size of the particles, as well as on the form and magnitude of the interaction between them.

Structured dispersions may be defined as all states of dispersed systems which are more ordered than a random distribution of single, free-moving particles and which fill the whole volume space of the dsipersion under consideration.

The term "more ordered" should not be exclusively interpreted in crystallographic terms, where "order" implies a periodic distribution of the particles (atoms, or molecules) in space. With coagulation structures, the word "ordered" may simply imply, for example, the fact that the particles, which initially were randomly distributed, without contacts, and free moving, are now fixed in their positions as a result of the coagulation. The mean random assay of the particles does not change perceptibly, if the particle concentration is sufficiently high. Hence, it is not always possible to "see" (for example by light scattering measurements) the sol-to-gel transition, in spite of the drastic change in rheological behavior at this point.

Figure 4.1 illustrates a simple two-dimensional random assay of particles, with (a) and without (b) coagulation. "Coagulation" implies that all particles within a certain separation (in this case smaller than one particle diameter) are physically "bonded". The structure shown in this case represents one "snapshot" of the randomly moving particles. Coagulation, and hence "ordering", has simply occurred by virtue of the particles becoming trapped in these positions. In other cases, where the particle concentration is too small to allow formation of structures like that shown in Figure 4.1, chains or "strings" of particles may be formed, which in turn may build a network throughout the bulk. To counteract the external forces (e.g., Brownian motion or non-directioned shear forces, induced by convection currents), which tend to disrupt any structure in a dispersion, internal forces are required which can initiate and stabilize

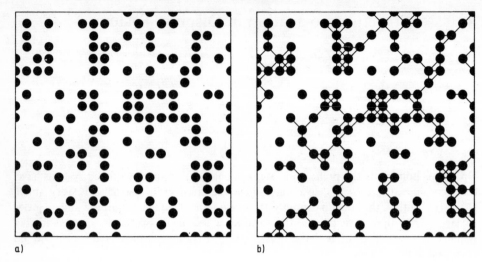

a)   b)

Figure 4.1
Formation of a coagulation structure (sol—gel inversion). (a) Snapshot of a random distribution of particles before coagulation; (b) after coagulation (all particles within distances smaller than one particle diameter are connected by bonds).

the structure. These internal forces may be either attractive or repulsive as we shall see. Alternatively, the action of external directed fields (e.g., gravitation, electric, magnetic, or shear fields) may promote structure formation.

Structured particulate systems are a frequent occurrence in daily life. To illustrate this, one may mention building materials (cement, mortar, concrete, clays), foodstuffs [2, 3] (ketchup, jellies, sauces, doughs), cosmetics, and pharmaceuticals (creams, emulsions, lotions, pastes). In biology also there are many examples of structured dispersions.

In order to understand the conditions under which structure formation takes place, a number of factors need to be considered. Particle size and shape, or better, the distribution of both, may in many cases be evaluated by optical or electron microscopy. Other, more advanced methods are based on small-angle light scattering from single particles in a flow ultramiscroscope [4—6] or photon correlation spectroscopy and similar techniques [7—11].

The concentration of the disperse phase is most usefully defined in terms of the volume fraction of the disperse phase, that is, the ratio of the volume of the dispersed phase to the true volume of the dispersion:

$$\Phi = \frac{v_{\text{disperse phase}}}{v_{\text{dispersion}}} \tag{4.1}$$

Structured systems, as defined above, only arise if Φ is greater than some critical value (Φ_{crit}). One way of obtaining this parameter experimentally is the measurement of the relative volume of sediments:

$$v_{\text{s}} = \frac{\text{volume of sediment}}{\text{volume of disperse phase in the sediment}} \tag{4.2}$$

which is easily done by dispersing a known volume (mass/density) of the solid in the liquid and measuring the equilibrium volume of the sediment.[1]) With different quantities of solids one normally obtains the same V_s (depending to some extent on the mode of dispersion and on the relaxation time of the structure; see below). This method, of course, is possible only if the density of the particles is sufficiently different from that of the solvent. A necessary criterion for structure formation therefore is

$$\Phi \geqq \Phi_{crit} = 1/v_s \tag{4.3}$$

Structure-forming disperse systems have comparatively high values of v_s (much higher than for dense packed spheres), i.e., small values of Φ_{crit} (see Fig. 4.2). Extreme examples of such structured dispersions are some clay minerals and pyrogenic silica particles, which at volume fractions as low as a about $\Phi_{crit} = 0.02$ (2% v/v) form coherent structures which fill the whole bulk. The determination of the three-dimensional

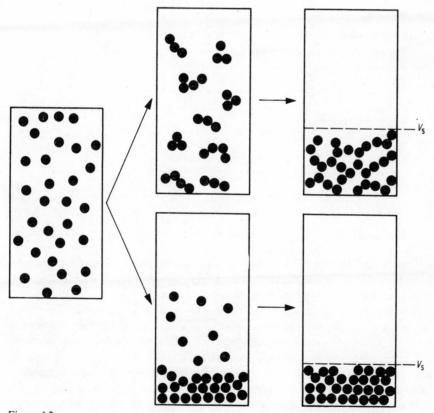

Figure 4.2
Sedimentation of stable and flocculated dispersions forming compact or loose structures.

---

[1]) It should be mentioned that these sediments are already compressed under the influence of their own weight [197].

distribution function of the particles is a much more difficult problem and will be dealt with later. Some ideas about the morphology of the structures are obtained from the analysis of the forces acting between the particles.

Basically, we have to distinguish between three types of structure formation:

— coagulation structures, due to attraction forces between the particles,
— periodic structures, due to repulsion forces, and
— dense packing of spheres.

The last type will not be considered in this book. Different types of dense packings have been dealt with extensively by Manegold [12]. They may form, if non- (or weakly) interacting particles sediment undisturbed, or if weak structures are destroyed by mechanical energy (e.g., by vibration). The volume fraction of solids in these dense packings is very high, as one may see from Table 4.1. Structures of the first type are the most important in practice and are therefore considered first here.

Table 4.1

Coordination number N_k and volume fraction Φ_{dp} for some dense packings of spheres

Type of packing	N_k	Φ_{dp}
Hexagonal	12	0.74
Body-centered cubic	8	0.68
Simple cubic	6	0.52
Tetrahedral	4	0.34
Random [59]	about 7.5	0.64

4.1.   Formation of Coagulation Structures by Attractive Forces

The spontaneous aggregation (flocculation or coagulation) of disperse phase particles has been the best studied process occurring in disperse systems [13]. Of great scientific and practical interest is the type of coagulation that gives rise to a solid-like spatial structure (a coagulation structure), due to the cohesion of particles, not into compact aggregates, but into chains and random networks of such chains. In many cases such coagulation structures develop by condensation of a new phase from a supersaturated solution. The precipitation rule of von Weimarn [14, 15] is one typical example. It states that: in producing an insoluble precipitate by a chemical reaction between concentrated solutions, one obtaines highly dispersed systems which often form a gel. An example would be the preparation of iron hydroxide sol from concentrated solutions of $FeCl_3$ and NH_4OH [16]. Another technically important example is the gel phase formed in the manufacture of polyvinyl alcohol by the reaction of polyvinyl acetate with methanol to give polyvinyl alcohol and methylacetate. If more than a critical number of acetate groups is exchanged for OH-groups, the polymeric molecules becomes insoluble in methanol and form small particles, which eventually form a stiff gel. Aggregation of the small particles by mechanical energy (stirring) leads to polyvinyl alcohol in powder form.

In general, colloidal systems may be generated by the condensation or dispersion of matter. Therefore coagulation structures may also be generated from small particles, created by deaggregating bigger ones, as has been shown for cellulose, nylon, and other materials [17, 18]. Coagulation structures display certain characteristic mechanical properties: a comparatively low strength, creep (even at very low shear stresses), structural viscosity, and high elasticity. These effects are the consequence of the thin liquid layers between the particles. Much more striking is the fact that coagulation structures (also called particulate gels), similar to gels built from macromulecules, can immobilize as much as 96% v/v solvent. The reason for this behavior is not clear as yet.

If actual contacts between particle surfaces occur, condensation structures are built, which, in contrast to coagulation structures, have a high strength, but low elasticity, and show no structural viscosity. However, in many practical disperse systems it is often difficult to decide which type of structure is present, because all intermediate states between these two extreme types are possible. In some disperse systems one observes a continous change from one type of structure to another, for example, in the hardening of cements. Initially, cement is a mixture of water, certain oxide particles, and finely grinded filler, which form a coagulation structure. The oxide particles gradually dissolve in water, yielding a solution which is supersaturated relative to the new phase (a product of the chemical interaction with water) and, ultimatively, a newly formed crystal hydrate grows from solution. Under certain conditions, these crystals form junction contacts, giving a "stonelike" crystalline structure. For a review of coagulation and condensation structures see refs. [19, 20].

There are five kinds of attraction energy, which may lead to the formation of coagulation structures:

1. electrostatic energy (V_{el})
2. dispersion (or van der Waals-Hamaker) energy (V_D)
3. hydrophobic interaction (V_h)
4. dipole interaction energy (V_{dipole})
5. hydrogen bonds (V_{H-H})

The first three are long range energies (up to some tens of nanometers), the other two only act at distances comparable to molecular dimensions (up to some nanometers). In most cases, two or more of the various kinds of energies act simultaneously.

Electrostatic attraction occurs if the surface potentials of the particles have opposite signs. The approximation of Hogg et al. [21] for the heterointeraction of spherical particles, i.e.,

$$V_{el} = \frac{\varepsilon \cdot a_1 \cdot a_2 (\psi_{01}^2 + \psi_{02}^2)}{4(a_1 + a_2)}$$

$$\times \left[\frac{2\psi_{01} \cdot \psi_{02}}{\psi_{01}^2 + \psi_{02}^2} \ln \left(\frac{1 + \exp(-\varkappa d)}{1 - \exp(-\varkappa d)} \right) + \ln(1 - \exp(-2\varkappa d)) \right] \qquad (4.4)$$

shows, in a quantitative manner, that V_{el} becomes negative (implying attraction), if the signs of the surface potentials (ψ_{01}, ψ_{02}) are different. A corrected version of this equation was developed by Healy et al., but the differences are small (see chapter 1). Exact numerically derived values are available in reference [22].

Opposite signs of the surface potentials may arise not only in mixtures of different particles (for example, in the coagulation of a positively charged sol with a negatively charged one [23—25] but also on different areas of the same particles, as often occurs, for example, with clay minerals (see section 4.1.1.).

The most important contribution to the formation of coagulation structures comes from the van der Waals-Hamaker or dispersion energy. This type of interaction energy has been described extensively in the first chapters of this book. It is useful to recall here an approximate equation for the interaction between spheres:

$$V_D = \frac{A^* \cdot a}{12d} \qquad (4.5)$$

with $A^* = (A_1^{1/2} - A_0^{1/2})^2$; A_0, A_1 are the Hamaker constants for the liquid medium and the particles, respectively; a is the radius of the particles; d is the minimum distance between the particle surfaces. Between similar particles V_D is always negative (i.e., attractive). Only in some special two-component dispersions can one observe repulsive dispersion energies between dissimilar particles [26, 27, 194]. For an understanding of the coagulation structures formed by nonspherical particles, to be discussed later, it is important to be able to calculate their dispersion energy; this is usually anisotrope. For example, as Kihara and Honda [28] have shown, the interaction energy between prolate ellipsoidal particles is at maximum if their long axes are collinear.

If the interaction between the solvent molecules is much greater than that between the solvent molecules and the surface, one can observe the so called "lyophobic" interactions. It is best known as the hydrophobic interaction when water is the solvent. Hydrophobic surfaces tend to be rejected from the water structure, in the same way that hydrophobic molecules are or the way in which the hydrophobic parts of amphiphilic substances are forced to association (i.e., to micellation) [29]. This effect tends to promote coagulation between particles. Very loose structures are obtained if only parts of the surface of the particles are hydrophobic (i.e., with mosaiclike surfaces). The occurrence of the hydrophobic forces is associated only with the surface properties of the particles and not their bulk properties. Thus the coagulation induced in certain dispersions by the thermal or chemical dehydration of surface (adsorbed) groups illustrates this point. If the dehydrated surface becomes hydrophobic, the particles aggregate under the influence of the hydrophobic interaction. Other ways of inducing hydrophobic interactions are to change either the nature of the solvent (by mixing with another solvent) or the nature of the surface (e.g., by adsorbing amphiphilic molecules, as in flotation). In flotation the adsorption of a surface active agent, with its hydrophobic part directed towards the water, leads to particle agglomeration both between themselves and with air bubbles.

A simple model for the hydrophobic interaction between particles has been presented by Cecil [30]. He coated the surface of glass beads (0.2 mm diameter) with a layer of

CH$_3$-groups, by reaction with dichlorodimethylsilane. The transfer of the beads from an aggregating to a nonaggregating solvent involves a change both in the area of contact with the liquid (A) and in the interfacial tension (γ). The change in free energy may be expressed as

$$\Delta G = \gamma \cdot \Delta A + A \cdot \Delta\gamma - T \cdot \Delta S \qquad (4.6)$$

where ΔA is the area of contact between aggregated beads, and ΔS is the change of entropy, mainly due to the change in water structure. ΔG may be negative, if ΔS is high enough.

Experiments including the shaking of hydrophobic beads in water—alcohol mixtures, with increasing alcohol content, revealed the following behavior. In pure water clusters of beads of about 50 particles are formed. At 35% v/v ethanol the clusters, formed in mixtures of lower alcohol content, are broken down to individual beads. The "changeover" concentrations for ethanol and other alcohols are shown in Table 4.2. Between 25 and 40 °C no temperature dependence was observed. Similar results were obtained using the method of creep measurement (see later and ref. [31]) to determine the fast elastic deformation modulus G_0 with hydrophilic silica particles (Aerosil 200), and hydrophobic (methylated) silica particles (Aerosil R 972) in ethanol—water mixtures (see Fig. 4.3). The modulus reflects the magnitude of interparticle interaction. With ethanol concentrations <40% v/v, the strong hydrophobic interaction of the hydrophobic R 972 particles made measurements impossible. With increasing ethanol content, the attraction between the particles gradually decreases. With the hydrophilic Aerosil 200 antagonistic behavior was observed. That is, with increasing concentration of the less polar solvent (ethanol), the attraction increased. This could suggest that there might also exist an oleophobic interaction, that is, an attraction between oleophobic particles in less polar solvents.

The first quantitative measurements of the net interaction, as a function of distance, between hydrophobic surfaces were reported recently by Israelachvili and Pashley [32, 207]. They measured the force of interaction between two mica sheets, supported on glass cylinders, which were hydrophobized by adsorption of cetyl trimethyl ammonium bromide. They demonstrated in these measurements that the hydrophobic interaction

Table 4.2

"Changeover" points for 0.2-mm methylated glass beads in different water—alcohol mixtures (after ref. [30])

Alcohol	Alcohol concentration in	
	% v/v	Molar ratio
Methanol	50 ± 6	0.44 ± 0.11
Ethanol	35 ± 5	0.18 ± 0.05
n-Propanol	19 ± 2	0.06 ± 0.01
n-Butanol	6.5 ± 0.5	0.014 ± 0.001

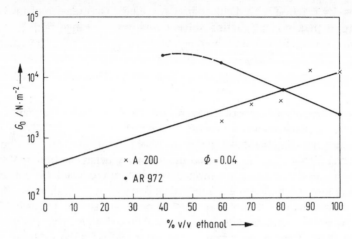

Figure 4.3
Influence of the ethanol concentration in mixtures with water on the modulus G_0 of 4% v/v dispersions of hydrophilic (A 200) and hydrophobic (AR 972) silica.

is a long range one, with an exponential decay, and is approximately one order of magnitude stronger than the dispersion interaction.

Attraction between dipoles is described by the equation [33]

$$V_{\text{dipole}} = -\frac{\mu_A \cdot \mu_B}{d^3}\; 2\cos\Theta_A \cos\Theta_B - \sin\Theta_A \sin\Theta_B \cos(\psi_A \cdot \psi_B) \quad (4.7)$$

where μ_A and μ_B are the dipole moments, d is the distance between the centers of the dipoles, Θ_A and Θ_B are the angles in inclination of the polar axes to the line of centers, and ψ_A and ψ_B are the angles subtended between the polar axes and perpendiculars passing through their centers.

Maximum attraction results if the dipoles are arranged in one plane, in a head-on-tail configuration ($\Theta_A = \Theta_B = \Theta$), i.e.,

$$V_{\text{dipole(max)}} = -\frac{2\mu_A \cdot \mu_B}{d^3} \quad (4.7\,a)$$

Hydrogen bonds exist between proton donor groups and proton acceptor groups which may be located on the same or on different molecules or particles [34]. Some characteristics of these groups are

— The hydrogen atom of a donor group must be bound to an electronegative atom, or, in other words, the donor group should have a partly ionic character.
— In the acceptor group readily displacable electrons are required. These may be the electrons of an electronegative atom or the π-electrons in unsaturated or aromatic systems.
— The hydrogen bond is a directed interaction. Its stability is at maximum if the axes of the donor and acceptor groups are collinear.

— The energy of a hydrogen bond is between 8 and 40 kJ $\cdot$ mol^{-1}, that is, higher than
the van der Waals interaction between molecules, but smaller than covalent chemical
bonds.

Some typical donor groups for structure formation are OH— (water, alcohols, carboxylic
acids, silanols) and NH= (amines, amides). Typical acceptor groups would be $=O\rangle$
(water, alcohols, ethers), $\rangle C=O\rangle$ (ketones, ethers, carboxylic acids), and π-electrons
(aromatic compounds). The best technique for detecting hydrogen bonds is infra-red
spectroscopy.

One typical example illustrating structure formation resulting from hydrogen bond
formation are the structures which may form in dispersion of silica particles in water or
when other strongly hydrogen-bonding molecules, such as ethylene glycol or its polymers
are present (see Fig. 4.4).

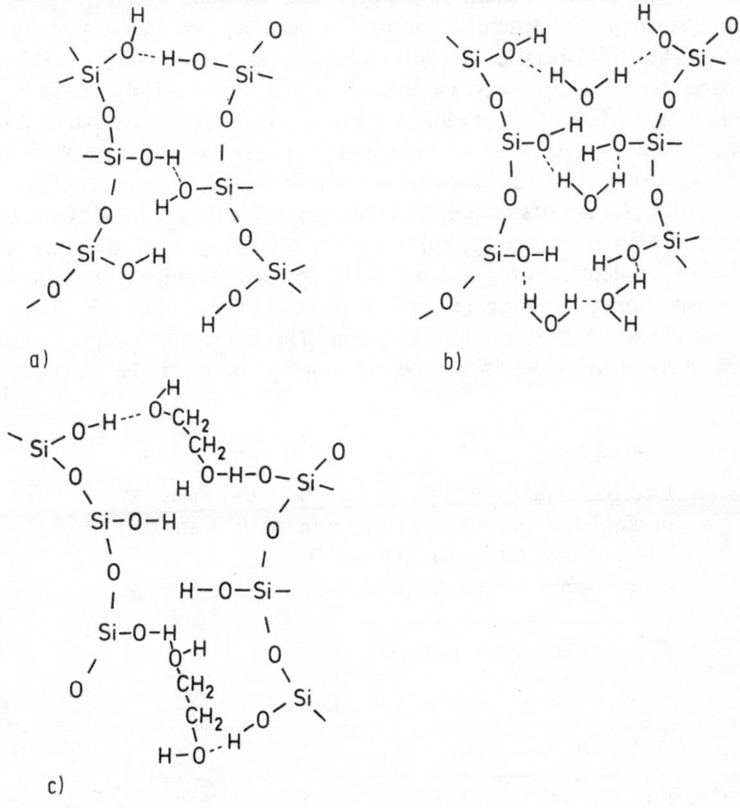

Figure 4.4
Examples for hydrogen bonding in silica dispersions in water: (a) silanol—silanol;
(b) silanol—water—silanol; (c) silanol—ethylenoxid—silanol.

4.1.1.   Coagulation Structures Formed by Anisotropic Particles

If the particles of a dispersion are anisotropic, they are quite likely to form coagulation structures. Anisotropy, of shape (anisometry) leading to structure formation is easily recognized (for example, the structures build by children, playing with a box of bricks or a pack of playing cards). Therefore a common opinion is that loose structures are only possible with anisometric particles. As will be seen later, however, such structures may also result with spherical particles. Gels forming from anisotropic crystallites are rarely discussed, exept for those derived from certain clay minerals (kaolinite, montmorillonite, bentonite) [50]. These have been used since ancient times by potters because of their ability to form plastic materials when mixed with water, which can be easily moulded or cast.

The alumosilicates crystallize in lattices with layers of $Al(OH)_2^+$ sandwiched between layers of $(Si_2O_5)^{2-}$. The particles of such materials are therefore expected to have a positive charge at their edges (where the positive Al-layers came to the surface) and a negative charge on their flat sides (this is a rather simplistic description of the crystal lattice structure of alumosilicates; for more details see ref. [50]).

Such anisotropy of charge distribution has been verified experimentally by Thiessen [51], who treated a dispersion of kaolinite with a negatively charged gold sol. With the aid of electron microscopy he was able to show that the negatively charged gold particles "decorated" the edges of the kaolinite crystals. In a more recent paper [52] values for the electrokinetic potentials of the faces and the edges have been reported as a function of pH value and electrolyte concentration (see Table 4.3).

The opposite charges of the faces and the edges leads to very loose "card-house" structures (see Fig. 4.5). Some typical electron micrographs of different concentrated "card-house" structures, obtained by the freeze-etch technique, may be found in ref. [20]. A dispersion containing just a few percent of the solid may yield a structure which fills the whole bulk of the dispersion. The minimum volume fraction required for such a structure depends on the relationship between the thickness δ and the

Table 4.3

Electrokinetic potentials of the edges and faces of kaolinite particles, as function of pH and NaCl concentration (in $m \cdot dm^{-3}$) (rearranged from ref. [52])

| pH | Electrokinetic potential in mV | | | |
| | Edge | | Face | |
	10^{-4}	10^{-1}	10^{-4}	10^{-1}
6	14	4	−54	−26
7	4	1	−54	−26
8	−14	−4	−54	−26

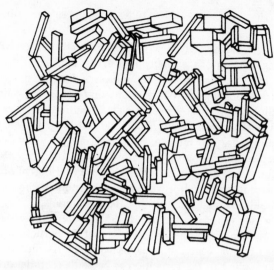

Figure 4.5
Card-house structure formed by platelike particles.

mean edge length a of the particles (which should be $\delta/a \ll 1$), and on the absolute value for a. Considering a symmetrical model based on squarelike plates, for the "card-house" structures, for the evaluation of φ_{crit}, one obtains

$$\Phi_{crit} \approx \frac{3\delta}{a + 3\delta} \tag{4.8}$$

For typical kaolinites (as cited by Kuhn [53]), critical volume fractions between 0.18 and 0.25 are obtained (see Table 4.4). For bentonite platelets, 1 μm in diameter and 2 unit layers thick (unit layer = 1 nm, particle thickness $\delta = 2(n - 1) + 1$; n is the number of unit layers) [50], one obtains from equation (4.8): $\Phi_{crit} = 0.009$, a value in the range of the experimentally observed one.

It can be readily shown that the opposite charges are in fact responsible for the stability of the "card-house" structures of clay minerals. If one increases the pH of the

Table 4.4
Φ_{crit} evaluated after equation (4.8) for some typical clay minerals

Type of clay	a in nm	δ in nm	Φ_{crit}
Kolloid Kaolin Chodau	400	30	0.18
Zettlitzer Kaolin	480	41	0.20
China clay No. 10	590	68	0.26
Schattenbacher Kaolin	1200	135	0.25
Bentonite	1000	3	0.009

suspension, the sign of the charges of the Al-layers changes (see Table 4.3) and a liquid dispersion results.

Some microcrystallite gels, derived from a macromolecular matrix by controlled chemical hydrolysis or degradation of the amorphous areas, have been investigated by Erdi and co-workers [17, 18]. They discuss the preparation and the rheological behavior of the gels formed from the following microcrystals: nylon-66 (spheres, 5 nm diameter, in water), regenerated cellulose (spheres, 5 nm, in water), natural wood pulp cellulose (rods, 100 nm long, in water), amylose (rods, 200—300 nm long, in water), collagen (spheres and elongated ellipsoids, 10—150 nm, in water), chrysolite asbestos (fibrilles 300—500 nm long, in tricresyl phosphate). The nature of the forces between particles is not dicussed by the authors, but it seems clear that they are of different origins for the different materials. In a recent paper [54] it was shown that this method of degrading synthetic fibers can profitably be used for recycling these materials.

In another gel forming sol, viz., vanadin pentoxide sol, the rodlike particles have a large permanent dipole moment. It has been shown [1] that the particles arrange themselves in long chains, head-to-tail, which may themselves undergo further structuring (at higher concentrations) leading to the formation of tactoids (parallel orientation of the chains).

The formation of a coagulation structure, due only to the anisotropic shape of the particles concerned, has been demonstrated by Okamoto and Hachisu [55]. For platelike gold particles there exists a pronounced secondary minimum in the interaction energy, when the platelets are oriented with their faces together. In the course of coagulation they at first form chains (with separation distances between the surfaces of the plates up to 100 nm!), but the chains later align in parallel rows (similar to the tactoids, mentioned above).

4.1.2.   Structures in Dispersions of Spherical Particles

For a long time it was the common opinion that coagulation structures are only possible if the primary particles are anisotropic, in spite of the fact that in the older literature it had been demonstrated that structure formation is possible in supensions of spherical particles [35, 36].

It was Usher [36] who was the first to draw attention to the distributions of the diffuse electric double layer around single particles and around aggregates, as being an important factor in the gelling (that is, the formation of coagulation structure) of inorganic sols of spherical particles. He showed from model calculations that the potential barrier around a doublet, at a given concentration of electrolyte, is anisotropic. The height of this barrier is smaller in the direction of the long axis than perpendicular to this axis. Further aggregation beyond the doublet stage leads to the formation of chains or branched structures. Usher carried out experiments with spherical particles of gambodge (the resin from the tropical tree Garcinia Hamburyi) having a mean radius of 400 nm. With the aid of a light microscope he observed the formation of aggregates, as a function of the nature and the concentration of added electrolyte.

With 0.05 mol $\cdot$ dm^{-3} NaCl the slow formation of doublets was observed. With 0.1 mol $\cdot$ dm^{-3} NaCl he saw doublets, triplets, and chains with 4—5 particles, which were mainly branched. With 1.0 mol $\cdot$ dm^{-3} NaCl a fully branched network of particle chains resulted, which he describes as follows: "... a common type of structure was a collection of loosely packed and comparatively large aggregates united by much more slender filaments ...".[1]) Usher also gave some hints, as to the critical volume fraction for structure formation (Φ_{crit}) which does not simply depend on the total volume of the solid, but on the product of the particle radius and the number of particles. From his experiments with gambodge particles ($\Phi_{crit} = 0.14$), and also with the much smaller particles of cadmium sulfide ($\Phi_{crit} = = 0.015$), he confirmed his predictions.

This older work has been referred to in such detail because it seems necessary to point out that the formation of gels from sols at the required electrolyte concentration was first described, both theoretically and experimentally, 50 years ago. The reader of recent papers on structure formation may not always receive this impression (see, for example refs. [38—40]). Usher also showed numerically that, in an encounter of a doublet and a singlet, the probability for adhesion along the long axis of the doublet is greater than at the side.

Hoffmann [41] in 1943 made similar calculations and Rees [42] in 1951 elaborated this idea with the aid of the DLVO theory. By calculating the lines of equal potential around a two-particle-aggregate, as a function of the electrolyte concentration, he was able to show that the height of the energy barrier at the end of the particle aggregate is lower than at the sides. Some 20 years later, Thomas and McCorcle [40] published similar calculations, using more exact (but more complicated) equations, obtaining the same results[2]).

During the coagulation process the chains of particles become longer. However, the probability of contact is related not only to the height of the energy barrier but also to the area of contact. Therefore, since with increasing chain length the area of the sides increases (the two end areas are, of course, constant), the probability of contact at the sides of the chains increases. At a critical chain length L the probabilities both for side and end become equal and the formation of chains is no longer favored. This critical chain length characterizes the length of the linear fragments in a coagulation structure formed by sedimentation of aggregates.

Values for $L/2a$ (a is the particle radius) ranging from 2 to 12 have been calculated [43, 44] for various values of the surface potential, Hamaker constant, particle size, and Debye-Hückel length. L increases with increasing double layer thickness and increasing height of the energy barrier. An increase of the Hamaker constant leads to a decrease of the barrier height and, therefore, to a decrease in L. Since the interparticle interaction energy in the DLVO theory is independent of the particle concentration, the average length of the chains formed during slow, irreversible

---

[1]) Is it just a curious fact that nearly the same words are used by an astronomer to characterize the structure of the universe? "... when distant galaxies are plotted on a map of the sky, the result is a striking tangle of clumps, and filaments, and voids ..." [37].

[2]) It seems to be a law of science that about every 20 years the older works are repeated by the next generation.

coagulation should also be independent of the particle concentration. However, for high particle concentrations this assumption is not correct [49]; the chain length above a certain particle concentration appears to become smaller. In concentrated dispersions, the spatial distribution of the particles is of fundamental importance, because in this case, structure results from contact of nearest neighbors and not by diffusion.

Chainlike aggregates in coagulated solid/liquid dispersions have been observed experimentally in many different systems. However, it is not always clear whether they are artefacts associated with the preparation of the samples for investigation by microscopy. Usher [36], as mentioned above, directly observed chains in dispersions of gambodge, whereas Thomas and McCorcle [40] (thoria sols), Strenge [45] (colloidal nickel particles in hydrocarbons), Bargeman et al. [16] (iron oxidhydrate sols) did so after different methods of sample preparation for electron microscopy. Hachisu [46] has published pictures of ordering in nonaqueous systems near the bottom of the container; in the early stages chains of 5—7 particles are observed to form. Rehbinder and Vlodavets [47] have studies the formation of interesting types of aggregate structures arising from the precipitation process in aqueous polymer solutions. An insoluble polymer formed as a result of the partial acetalization of polyvinyl alcohol with formaldehyde in an acidified aqueous solution. Supersaturation leads to the precipitation from solution of very fine particles of polymer. Subsequently, these particles form entangled chains. Further acetalization and drying of such structures leads to elastic materials (used, for example, as leather substitute).

Microscopic studies of the formation of particle aggregates in very slow laminar shear fields have been made by Vadas and co-workers [38, 48]. They were able to show that for, polystyrene latex (diameters from 0.92 to 3.75 µm and particle concentrations from 10^8 to 10^9 cm^{-3}), with increasing sodium chloride concentration there is a marked tendency for the formation of chainlike units. For example, at 0.3 mole $\cdot$ dm^{-3} (i.e., close to the critical coagulation concentration) 68% of the triplets and 91% of the quadruplets occur as linear or nonlinear chains.

It was stated earlier that the structure of a gel results from the coagulation of primary spherical particles is a tangled network of particle chains. There are four main ways of obtaining some insight into the nature of such gels:

1. direct modeling of the structures, using large particles,
2. computer simulation of aggregation and sedimentation,
3. using mathematical models for branching systems (percolation theory),
4. experimental studies on gels, such as their rheological behavior, solvent diffusion, and electric conductance.

4.1.2.1.   Direct Modeling of Structures

Our measurements on dispersions of pyrogenic silica (Aerosil 200) have shown that structuring occurs at volume fractions in excess of 1%—2% v/v (in water $\Phi_{crit} = 0.018$, in 3 mole $\cdot$ dm^{-3} KCl solution $\Phi_{crit} = 0.012$, and in n-decane $\Phi_{crit} = 0.012$).

At 4% v/v all these dispersions resemble a "paste". Attempts were made to model the structure at this concentration using larger particles. The original silica particles had a mean diameter of 12 nm; the model particles (foamed polystyrene spheres) were 6 mm in diameter, i.e., a factor of $5 \cdot 10^5$ greater. The particles were contained in a $10 \times 10 \times 10$ cm³ cube; this corresponds to a real volume of $V_m = 200^3$ nm³. The number, n, of particles per unit volume in a dispersion is given by

$$n = \frac{6\Phi}{\pi D^3} \tag{4.9}$$

for $\Phi = 0.04$ and $D = 12$ nm, $n = 4{,}42 \cdot 10^{16}$ cm⁻³. Hence, there are 354 particles in the volume element that is to be modeled. Several different structures may be proposed. For spherical particles, three possible configurations are shown in Figure 4.6 [36, 39, 56, 57]: (a) All the particles are arranged in a statistically distributed network of chains; or (b) there are some denser flocs or clusters of particles, bonded by longer chains, or (c) each sphere is surrounded by a sheet of solvent molecules such that these "effective" particles close pack and fill the whole volume. (This arrangement actually belongs to the group of periodic structures, to be discussed in detail later.)

It was only possible to fill the whole volume space of the model randomly with 354 particles using linear chains with a maximum of about five particles per chain. A detailed study of an Aerosil dispersion using ultracentrifugation and viscosimetry, after the application of ultrasound energy, has shown [58] that complete deaggregation to primary particles ($D = 12$ nm) does not occur, but, rather to small flocs containing an average of about 20 particles, which had been fused together during original formation. The volume fraction of the particles in these aggregates is $\Phi_A = 0.40$, and the diameter of the aggregate is $D_A = 48$ nm. Assuming these denser, primary flocs to be connected together by longer chains, it proved to be very difficult to fill the whole volume space of the model.[1]) These model do raise some interesting new questions, for example:

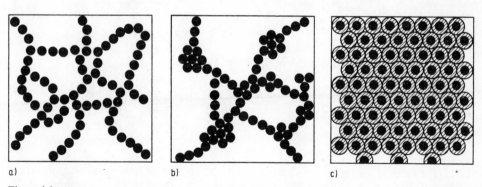

a)   b)   c)

Figure 4.6
Three possible types of structure created in dispersions of spherical particles: (a) simple network of chains; (b) denser flocs, connected by chains; (c) dense packing of spheres surrounded by a thick immobile layer of solvent or adsorbed species.

---

[1]) For technical reasons, unfortunately, it is not possible to reproduce here any photographs of the scaled-up models. It would, however, be an interesting pastime for the reader to construct and consider similar models for himself.

why is the solvent (which, in the model, is made up of particles the size of fine sand) fully immobilized?

Structure c is unlikely to occur for silica dispersions in hydrocarbons. In Table 4.5 some data are presented for the thickness δ of the solvation layers necessary to obtain the packing densities required. The thickness is given by the equation

$$\delta = \frac{D}{2}\left(\sqrt[3]{\frac{\Phi_{dp}}{\Phi}} - 1 \right) \tag{4.10}$$

For $D = 12$ nm and $\Phi = 0.04$ the corresponding δ values are given in Table 4.5 for various types of packing. A thickness of 8—10 nm for a rigid layer of adsorbed, structured solvent (decane) is not very probable.

Table 4.5

Necessary thickness of an adsorbed layer, leading to the dense packing of spheres ($D = 12$ nm, $\Phi = 0.04$)

Type of packing		
Hexagonal	0.741	9.9
Body centered cubic	0.686	9.5
Simple cubic	0.524	8.1
Tetrahedral	0.340	6.2
Random [59]	0.64	9.1

Summarizing the results obtained from the direct-modeling experiments, it may be concluded that, for the very loose structures found experimentally in the case of the fumed silica dispersions, a network of particle chains, in which some denser flocs are embedded, must be assumed.

4.1.2.2.   Computer Simulation of Structure Formation

The fast calculation possibilities of modern computers give us a new tool for the investigation of "structure" formation in colloidal dispersions. The formation of flocs or sediments can be simulated by a computer program in two different modes:
1. by randomly putting and/or moving particles in a given volume (Monte Carlo simulation technique) or
2. by the method of molecular dynamics, i.e., by the simultaneous calculation of the exact equations of motion for a large number of particles in a given volume.

In both cases special conditions for the flocculation process have to be assumed. These conditions may be empirical (in the manner: "1/3 of all encounters lead to adhesion"), or exact (real distribution of interaction potential around particles, kinetic, and hydrodynamic conditions).

In the review paper of Medalia [60] it is mainly the first type of computations that is extensively covered. The main papers referred to in that paper will be summarized here also. The first person who tried to use computer simulation for understanding floc and sediment formation was Marjorie J. Vold. In her first paper, on the formation of sediments [61], she assumed permanent adhesion on contact of the particles. Each particle was sequentially placed in a random position at the surface and allowed to sediment until it contacted the bottom of the container or another particle. The storage capability of the computer used permitted simulation of only 160 particles. In the central region of the computed sediment, which was assumed to be representive of a real sediment, a particle volume fraction of 0.12—0.13 was determined. In a second set of computations it was assumed that attraction forces operate over varying separations between the particles. A large decrease in the volume fraction was found with increasing range of attraction. If, for example, cohesion is assumed to occur at a separation distance of $2a$ (a is the particle radius), Φ_{crit} decreases from 0.13 (direct contact) to about 0.06. This is in qualitative agreement with the observation that strongly flocculated dispersions form very "loose" sediments.

In a later paper, Vold [62] introduced a factor that randomly determines the fraction of encounters that lead to cohesion. The volume fraction decreased smoothly on going from 0 to 100% cohesion probability. The next step was to model the influence of particle anisometry [63]. Rigid chains of particles are modeled as anisometric particles. For an increase in the number of particles per chain from 2 to 18, a steep decrease in the volume fraction in the sediment results. Other information may be obtained from simulations of this type, for example, the number of contacts between particles, or of the end-to-end, end-to-side, and side-to-side contacts between rodlike particles. Vold and others have concluded from a comparison between calculated and experimental values of these contact numbers, that end-to-end contacts (i.e., chain formation) are generally favored. (This is in good agreement with the experimental and theoretical results presented in section 4.1.2.)

Since sediments in flocculated systems are not formed from single particles or rigid chains, but by flocs, Vold also tried to simulate floc formation [64]. Again particles were sequentially generated, moved at random, and the structure of the resulting aggregates was registered. The general appearance of Vold's flocs is that of a central core, surrounded by long tentacles. This series of papers by Vold gives a good impression of how much new information may be obtained from relatively simple Monte Carlo computations, provided the conditions are defined exactly.

On the basis of Vold's calculations, Sutherland improved the method [65—70]. Taking the kinetics of coagulation, as presented by Smoluchowski, to formulate the conditions for floc formation, he generated very open flocs of low density with no clearly defined core. Smoluchowski kinetics takes into account the contact between single particles, single particles and higher aggregates, and higher aggregates with other flocs, but neglects the differences in interaction energies between particles of different size and shape. A further improvement in the software of Sutherland to take account of this would lead to simulated floc structures that would be a quite realistic representation of real flocs and sediments.

The other main approach to computer simulation is the method of molecular dynamics. This method requires a large computer memory because, for every particle in a large assembly, the exact equations of motion over small time intervals must be calculated and the positions stored. In addition, the effect of all neighbor particles on the motion of any given particle must be taken into account. Until now, only some preliminary calculations for colloidal systems have been published. Deutch and Oppenheimer [71] calculated the Brownian motion of several colloidal particles. Ermak and McLammon [72] have investigated the Brownian dynamics for larger assemblies introducing hydrodynamic interactions. Dickinson and co-workers [73] calculated the Brownian motion of flocs for two or three interacting particles again including hydrodynamic effects, and in another paper [200] Dickinson takes into account the theory of coagulation of concentrated dispersions. Reference [73] gives values for the lifetime of doublets and triplets, based on the DLVO theory. To go from this point to actual "structure" formation in large flocs is mainly a question of computer capacity,

A completely different method for the modeling of structure formation is the use of percolation theory. This was proposed by Makarov and Suško [74, 75], based on a general paper by Fisher and Essam [76]. A coagulation structure may be viewed as a stochastic random sample, because it consists of a great number of particles with random interactions. The location of any single particle in the structure is not predictable. The structure may be assumed to be a branched (treelike) one, or a network. In the simplest case the particles occupy random sites on a lattice. Each site can only accommodate one particle, and has a constant occupation probability factor P. A group of particles linked together by nearest neighbor "bonds" on adjacent lattice sites are said to form a "cluster". For real lattices of principal interest it is very difficult to formulate a direct theoretical approach leading to solutions in closed form. Appreciable insight may be obtained, however, by examing pseudolattices, such as the Bethe lattices [77] (i.e., infinite homogenous Cayley trees), and for example, the various triangular "cacti" which may be generated in a cascade process (see Fig. 4.7).

This is not the place for a detailed discussion of the methods of calculation. Rather, some of the important results will be given. Fisher and Essam [76] calculated

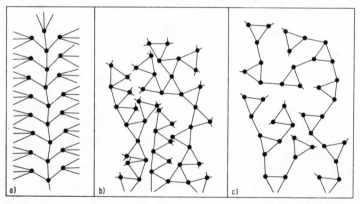

Figure 4.7

Simple pseudolattices: (a) Bethe lattice (Cayley tree) with a coordination number of 4; (b) simple cactus with coordination number of 4; (c) expanded cactus with coordination number of 3 (redrawn after ref. [76]).

for a simple Cayley tree, with a coordination number N_k, the critical probability, P_c, above which unbounded infinite clusters (i.e., structures) are formed,

$$P_c = \frac{1}{N_k - 1} \tag{4.11}$$

Below P_c, simple finite clusters (i.e., flocs) result, which grow to a maximum size at P_c. Makarov and Suško [74] tried to find the connection between the critical concentration for structure formation and the probabilities given by the percolation theory. They take the critical concentration for structure formation Φ_{crit} as the concentration at P_c, and the concentration where the structure attains a "nearly solidlike" behavior (with diminishing thixotropy) Φ_∞ as that at $P = 1$. Then

$$\frac{\Phi_\infty}{\Phi_{crit}} = \frac{P}{P_c} = N_k - 1 \tag{4.12}$$

or

$$N_k = \frac{\Phi_\infty}{\Phi_{crit}} + 1 \tag{4.13}$$

For Aerosil 200 in decane, taking $\Phi_{crit} = 0.012$ (see sect. 4.1.2.1.) and Φ_∞ about 0.03, then $N_k = 3.5$. Direct modeling has shown that only for $N_k < 3$ is the construction of a volume filling structure possible. Makarov and Suško give experimental values (for methylated Aerosil in decane: $\Phi_{crit} = 0.04$ and $\Phi_\infty = 0.08$. With these figures they found $N_k = 3$. For $N_k = 3$, the minimal possible volume fraction of equal spheres is 0.056 [2], i.e., higher than that calculated by the authors. The reasons for these differences between the theory and the experiments are not yet clear, but the process of building up a coagulation structure resembles in many ways the typical pattern for a percolation process: the building of single flocs (finite clusters); a critical concentration for the creation of a volume-filling structure (i.e., infinite clusters); a rise of the strength of the structure with increasing concentration. The biggest difficulty lies in the determination of the value for Φ_∞, which is directly related to the number of contacts, N_k, and which cannot be measured directly (i.e., by rheological methods) at the point $P = 1$, as has been done for conductivity measurements on resistor models [78] and microemulsions [79].

4.1.2.3.   Conclusions

Methods have been described for the theoretical modeling of coagulation structures built from spherical particles at low or medium particle concentrations (i.e., concentrations much less than that required for the dense packing of the spheres). From these approaches, together with results from rheological experiments, it is concluded that such structures are only possible if networks of chains containing 4—5 particles are formed. These chains may contain, at the interconnecting nodes, dense packed particle assemblages. It has been shown in many experimental studies [16, 80, 81] that such structures are indeed obtained for a variety of solid particles in different solvents.

Such structures in most cases may be reversibly ruptured by shearing (i.e., show thixo-tropic behavior) and they often exhibit an elastic response to deformation, independent of the viscosity of the dispersing medium.

At high particle concentration, the behavior is more difficult to interpret because

1. the flow of the suspending medium on deformation of the structure cannot be neglected; and
2. deformation is not only stretching or bending of chains, but also interchange of particles between lattice sites.

4.1.3.   Kinetics of Coagulation-Structure Formation

Very little experimental work has been published on the kinetics of structure for-mation. The main reason for this is the difficulty in explaining theoretically the time-dependent behavior of a structure in statu nascendi. In dilute dispersions, used in most experimental investigations of coagulation kinetics, the rate determining step is diffusion.

The theoretical basis of diffusion is well known, and hence the theoretical interpretation of such kinetic experiments is well established. The appropriate time scale for the experi-ments may be chosen simply by changing the particle concentration. With more concentrated systems, i.e., these in which volume-filling coagulation structures may form, the theoretical basis for the kinetics is not so well established. The time-scale for such systems varies from seconds to weeks or months. The few published papers, to be discussed here, are concerned with rheological methods for monitoring structure formation. With the technically very interesting system: 5.5 % w/w Aerosil 380 in a mixture of sulfuric acid (density 1.280 g $\cdot$ cm^{-3}) with 5 % phosphoric acid (density 1.165 g $\cdot$ cm^{-3}) (a gelled electrolyte used in lead accumulators), Aguf and Orkina [82] measured the yield stress as a function of time. They used a capillary viscosimeter with variable pressure, at temperatures ranging from 25 to 45 °C. As may be seen from Figure 4.8, at ambient temperature structure is formed after about 5 h. Increasing the temperature results in somewhat shorter formation times, but the risk of syneresis (an undesirable process in many technologies) also increases with increasing temperature. Aguf and Or-kina found a pseudomolecular, second order reaction rate, i.e.,

$$\frac{dn}{dt} = k \cdot n \cdot (n_\infty - n) \tag{4.14}$$

n is the number of particles in the structure at time t, n_∞ is the number of particles in the equilibrium structure. Using equ. (4.63) for the calculation of the interaction energies from the elastic deformation modulus (see section 4.4.3.), they found the following values for the rate constant: $k_{25°} = 0.9 \cdot 10^{-19}$; $k_{35°} = 1.2 \cdot 10^{-19}$; $k_{45°} = 3 \cdot 10^{-19}$ cm^3 $\cdot$ s^{-1}. The activation energy for the structure-forming process was calculated to be ~ 38 kJ $\cdot$ mol^{-1}.

Erdi, Cruz, and Battista, in their previously cited paper on polymeric microcrystalline gels [17], also used the yield stress (measured as the maximal stress, developing at a constant, small rate of deformation) to characterize the kinetics of structure for-mation. As may be seen from Figure 4.9 the times needed to form an equilibrium

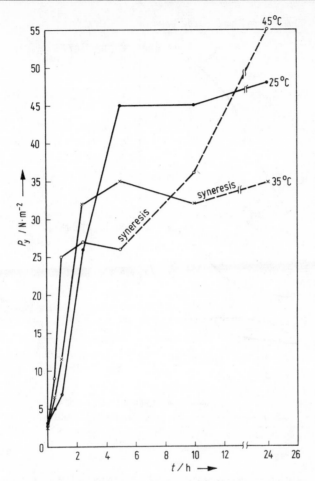

Figure 4.8
Yield stress of silica dispersions in sulphuric acid at different temperatures as a function of time after preparation [82].

structure vary from 10 minutes (for cellulose gel) to more then one week (for nylon or amylose).

With the aid of an analytical ultracentrifuge the present authors tried to monitor the rebuilding of structure in an Aerosil 200 (4% v/v in water) dispersion, following a virtually complete dispersion by ultrasonication [58]. There were great experimental difficulties: the samples had to be aged in the measuring cells of the ultracentrifuge and had to be transported to, and accelerated in, the centrifuge without disturbing the structure. Hence, only qualitative results were obtained, which showed that the reformation of the structure requires about 6 weeks. (After 7 days no significant change was detected; after 22 days the presence of a detectable structure was found; after 6 weeks the structure resembled the original state, prior to ultrasonication.)

10*

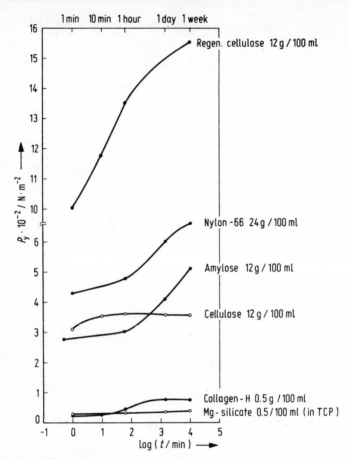

Figure 4.9
Yield stress of different microcrystalline materials in water (in TCP = tricresyle phosphate) as function of time after preparation [17].

Similar times for the formation of structures in Aerosil—water dispersions have been reported [196].

Structure formation, and thixotropic recovery, have also been monitored by Mewis [83] and Ur'ev [20] over time intervals of about 1 to 100 s using oscillatory measurements parallel or pendicular to the shear plane. Mewis and co-workers [84] also used an elegant method for monitoring structural changes by measuring the complex dielectric constant under shear conditions. Experiments on a dispersion of 7.5% v/v carbon black in mineral oil showed the usefulness of this method. One can "see" changes in the dielectric constant at short times (less than 10 s) after stopping the flow, and also up to 20 h. The reported measurements seem to show that, in this particular system, the reconstruction of the network is not completed after 20 h (see Fig. 4.10).

It follows from these few experimental studies that structure formation is a very complex process, governed by the extent of destruction (the prehistory), by the random internal (Brownian) and external (vibrations, temperature fluctuations) disturbances, and by the shape and surface properties of the particles. Much more work on the kinetics of structure formation has to be done. Indeed, this may be one of the best ways of gaining further insight into the nature of the structures that form in real systems.

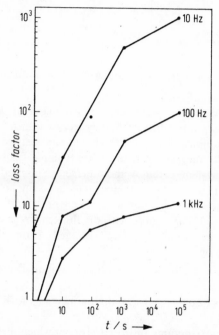

Figure 4.10
Loss factor of a 7.5 % v/v carbon black dispersion at different frequencies as function of the time after shearing at $\dot{\gamma} = 26.2 \, \mathrm{s}^{-1}$ [83].

4.2.   Periodic Structures Formed by Repulsive Forces

The term "periodic structures" was first used as the title of a monograph by Efremov [86, 87], which was concerned with all topics of structure formation. The term "periodic structures" will be used here, as stated before, only for real periodic systems, that is, for those structures in which every particle is the same distance from its nearest neighbors, at least in two dimensions. Such a structure is only possible if repulsive forces act between the particles. In other systems, also classed by Efremov as "periodic" (for example, thixotropic gels and pastes) the particles are only orientated over short distances in one dimension (usually as chain fragments). The general picture in these types of structure, however, is that of a random distribution.

To a first approximation, the nature of the repulsive forces is not important. They on every occasion give way to an energy minimum for one particle between its

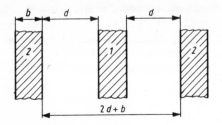

Figure 4.11
Model for the interaction of a semi-infinite plate (*1*) with two fixed plates (*2*) [86]. ·

nearest neighbors. This is best illustrated by using semi-infinite plates of thickness b (for spherical particles one has to use a three-dimensional approach, which leads to an analogous result which is not easy to illustrate). Since the interaction between platelets depends on their size and on the distance between them, the volume fraction of the plates is not a useful variable for this system. Efremov and others have inserted the use of the variable $(2d + b)/b$ (see Fig. 4.11), which may be interpreted as being analogous to the inverse of the particle volume fraction. If the interaction energy of a particle with its two neighbors is plotted as a function of $(2d + b)/b$, with the thickness b as a variable, then results such as those shown in Figure 4.12 are obtained. From this figure the following observations may be drawn:

(i)   in dilute systems ($d/b = 10$), the depth of the energy minimum decreases with increasing platelet thickness,
(ii) in more concentrated dispersions ($d/b = 3$), the minimum increases with increasing platelet thickness.

This lead to a division of periodic structures into two groups, which also differ in their behavior in practice: short-range (concentrated) and long-range (diluted) periodic structures. Both types have in common that the electrical double layers overlap: for the concentrated systems because of the high number of charged particles; for the diluted systems because of the very low concentration of electrolyte (corresponding to high r_D values) [198].

4.2.1.   Short-Range Periodic Structures

In concentrated systems of spherical particles, the repulsive forces act as if the particles have an effective radius that is equal to half the distance from the particle center to the minimum. These "effective" particles form close-packed arrangements that are highly ordered and periodic. If the volume fraction of the "naked" particles Φ and the mode of packing (regular packing with 8 or 12 nearest neighbors) is known, than one can easily calculate the thickness of the adsorption layer δ or of the effective thickness of the diffuse part of the electric double layer, r'_D,

$$\delta = \frac{d}{2} = r'_D = \frac{D}{2}\left[\sqrt[3]{\frac{\Phi_{dp}}{\Phi}} - 1\right] \tag{4.15}$$

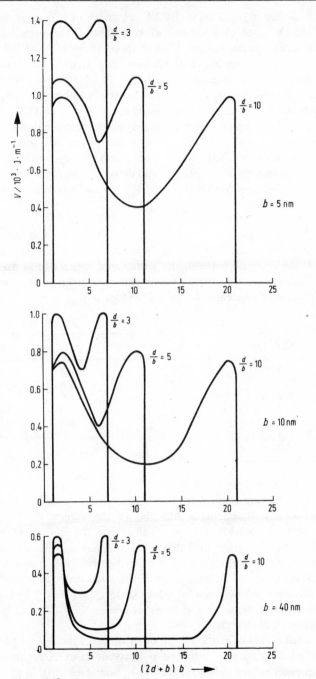

Figure 4.12
Interaction energy for particles like Fig. 4.11 for different particle thicknesses b and concentrations d/b [86].

with D the particle diameter [see section 4.1.2.1., equation (4.10)]. r_D' is not equal to r_D (the inverse of the Debye-Hückel parameter), because r_D is the (arbitrarily defined) distance at which the electrostatic potential has dropped to $1/e$ of its surface (or Stern) value. Interaction of the overlapping double layers may actually be observed up to $5r_D$, at low electrolyte concentrations. Φ_{dp} in equation (4.15) is the volume fraction of close-packed spheres in regular array, with a coordination number of 12 ($\Phi_{dp} = 0.741$, hexagonal close packing) or a coordination number of 8 ($\Phi_{dp} = 0.686$, cubic close packing). In diluted and moderately concentrated systems, hexagonal close packing is generally observed experimentally [88, 89]. This is due to the fact that the particles may easily reach the sites of their lowest interaction energy, without disturbing each other. At higher concentrations some "crystal defects" may occur. In this case, the random packing, as discussed by Mason [59] with $\Phi_{dp}' = 0.64$, is the most common.

Table 4.6

Particle distance and thickness of the electrical double layer
a) Half the distance between the surfaces of spheres ($d/2$) as a function of the volume fraction for particles with $D = 200$ nm, b) r_D and $1.9 \cdot r_D$ as a function of concentration of 1,1 electrolyte (c_{el})

a)			b)			
Φ	$d/2$ in nm		c_{el} in mol·dm^{-3}	r_D in nm	$1.9r_D$ in nm
0.10	95.0		10^{-4}	30.4	57.8
0.20	54.7		10^{-3}	9.6	18.3
0.30	35.2		10^{-2}	3.0	5.8
0.40	22.8		10^{-1}	1.0	1.8
0.50	14.0		1	0.3	0.6
0.60	7.3				
0.70	1.9				

For particles of $D = 200$ nm (typical for many synthetic polymer lattices), equation (4.15) leads to the values of $d/2$, as a function of the volume fraction given in Table 4.6. These may be compared to the values for the inverse Debye-Hückel parameter r_D, which are also given in Table 4.6 for different concentrations of 1,1 electrolyte. Also given are the values $1.9\,r_D$, discussed in ref. [90]. It seems that both $d/2$ and r_D over certain ringes of Φ and c_{el} are of the same order of magnitude. This corresponds with the observed behavior of many lattices, i.e., they "thicken" if they are dialyzed (c_{el} decreases, r_D increases and becomes greater than $d/2$), and become less viscous on the addition of electrolyte [91—94].

With regular ordered arrays of particles one often observes the optical phenomenon of iridescence. For electrostatically stabilized polystyrene latex dispersions, Hachisu et al. [95] have given experimental values for the Φ—c_{el} boundaries within which ordered structures occur for particles with a radius of 85 nm, showing iridescence. This work was reexamined by Barnes et al. [90], who found that, for the order/disorder boundaries given by Hachisu et al. (0.1 < φ < 0.3), the effective radius of the particle is given by

$$a = a + \lambda \cdot r_D \tag{4.16}$$

with $\lambda(a) = 1.9 + \ln (a/85)$, a in nm. The onset of ordering occurs if the value of Φ of the "effective" spheres exceeds 0.50, and complete ordering is achieved if $\Phi = \Phi_{dp}$ = 0.74. Increasing particle size favors the disordered state, since, at a given Φ, the electrolyte concentration has to be reduced before ordering is observed. This behavior is not in accord with the theoretical prediction stated previously, which would suggest that with larger particles the deeper minimum should favor ordering. The boundaries given by Hachisu et al. are very dependent on the method of observation. As will be seen later, structures are possible in 100 times more diluted systems.

For much smaller particles (compared to the latex particles) of SiO_2 and CeO_2 (= 6 to 80 nm), Ramsay [96] found repulsion structures occurred between 15% and 40% w/w (i.e., for SiO_2: 7%—18% v/v or $d/2$ = 3.6—24 nm, typical distances for the extension of the electrical double layer). This type of repulsion structure obtained with silica particles corresponds to the first step in the natural formation of the mineral opal. Precious opals consist of a regular array of ordered silica spheres of uniform size (100—500 nm diameter) which are set in an amorphous silica matrix [97, 98]. Attempts have been made to make synthetic opals [99], which are claimed to be indistinguishable from natural ones. Recently an interesting type of opal was analyzed which was built up from two interpenetrating lattices of spheres of two different sizes (180 and 100 nm diameter) [100]. The authors calculated that such mixed lattices are possible if the sum of the interaction energies between two large particles (V_{AA}) and between two small particles (V_{BB}) is larger than twice the interaction energy between larger and smaller particles (V_{AB}), i.e.,

$$V_{AA} + V_{BB} > 2V_{AB} \tag{4.17}$$

Hachisu and Yoshimura [101] were later able to model such crystalline superstructures with binary mixtures of latex spheres (550 and 270 or 310 nm in diameter). They found several types of structures, some of which resemble that of metallic alloys (Na $\cdot$ Zn_{13}, Ca $\cdot$ Cu_5). They concluded from this work that the repulsive forces have to be accompanied by some entropic effect to create such interpenetrating two-phase structures.

4.2.2.   Long-Range Periodic Structures

At very low electrolyte concentrations latices are found with distances between the particles up to 10 particle diameters. Such very loose structures are of great interest in crystal physics, because they reveal many features of crystals on an enormously magnified scale. Repulsive ordering is the central feature of the Wiegner crystal [117], based on a model of electrons moving in a uniform background of positive charge. The macroscopic model from polystyrene latex shows intense Bragg diffraction of visible light. Crandall and Williams [121, 122] used spheres 100 nm in diameter with 2000 electron charges per particle (that is about 1.02 $\mu C \cdot cm^{-2}$), to measure the Young modulus E for such "crystals" which form in aqueous suspensions. A very low ion concentration has been obtained using mixed ion exchanging resins. They measured the lattice constant, $a(z)$, as a function of the height, z, in the test tube from the Bragg diffraction peaks of the light of a helium—neon laser, $a(z)$ decreases uniformly from the top to the bottom of the cell. To confirm that this effect was actually due to gravitational compression of the lattice, they made the same measurements in D_2O also. The effective density $\Delta\varrho = (\varrho_p - \varrho_0)$ in equation (4.18) is $\Delta\varrho = 0.05$ g cm^{-3} in H_2O and $\Delta\varrho = -0.05$ g · cm^{-3} in D_2O. The modulus is calculated from the relation

$$\frac{a(z) - a(z_m)}{a(z_m)} = n \left(\frac{\pi D^3}{6}\right) \Delta\varrho \ g(z - z_m)/E \qquad (4.18)$$

z_m is the midplane in the column of the structure. The volume fractions in these experiments ranged from $5 \cdot 10^{-4}$ to $3.5 \cdot 10^{-3}$, the concentration of particles from $n = 10^{12}$ to 10^{13} cm^{-3}. The measured modulus E (of the order of 1 N · m^{-2}) increases with increasing particle concentration (i.e., with Φ), since the repulsion is larger when the particles are closer together. The modulus measured in the model structure is about one order of magnitude greater than that of an ideal gas ($E_{ges} = n \cdot k \cdot T$). A detailed investigation of the types of crystals and of the kinetics of their formation has been carried out by Clark et al. [123]. With particles of diameter $D = 109$ or 234 nm they obtained a cubic lattice constant of $a = 2200$ nm, i.e., about ten particle diameters. After shear-induced rupture of the structure, one first observes regions of hexagonal closed-packed crystallites build, which then later rearrange to body-centered cubic lattices.

Such extremely long-range repulsion forces between particles are at first glance difficult to understand, particularly in the light of the concentration boundaries for structure formation given by Hachisu et al. [95]. A simple calculation, however, using the accurate analytical expression given by Oshima, Healy, and White [124], shows that, under certain conditions (especially high surface potential and low electrolyte concentration), the repulsive interaction between two single particles is already high at separation distances as great as $d = 4$—5 times r_D. Assuming additive multiparticle interactions, one can readily account for the experimental results. A detailed study of the electrostatic repulsion "without a reservoir", i.e., in the case that the diffuse parts of the electrical double layers of all particles overlap, recently was given by Beresford-Smith et al. [198].

Anisometric particles may also form periodic structures. One example is the tobacco mosaic virus. These viruses are hollow cylinders, about 300 nm long with a diameter of 15 nm [125]. Above a critical concentration they form an iridescent gel which can be viewed as a periodic structure [116].

The ordered periodic structures are easier to handle theoretically than the random coagulation structures. As a result a large number of papers concerning ordered structure have been published, both on analytical theory and on computer modeling [102—117, 86, 87, 199]. These have succeeded in modeling periodic structure for the different system variables such as particle size and charge, Hamaker constant, temperature, and concentration.

The repulsive forces occurring in short-range periodic structures are not always electrostatic in origin: steric forces (i.e., thick solvated adsorption layers) are also possible.

One system of periodic structures, not mentioned up to now, are the liquid crystals. It is true that they are formed by molecules and not by particles, but the ordered subunits are of the size of colloids. The term "crystal" implies some element of periodicity in this type of structure, which occurs with certain types of organic molecules. These are molecules that associate to give molecular units that diffract light and give rise to turbidity, which is a characteristic of this state of matter. Liquid crystals are of great technical importance and their study has become a self-contained science. Only a few key papers will be cited here in which the reader will find further information on this subject [118—120].

4.3.   Formation of Structures under the Influence of External Forces

In the preceding sections those types of structure have been discussed that are formed as a direct consequence of interparticle forces. The presence of external force fields (electrical, magnetical, or mechanical) may not only lead to modification of these interparticle interaction forces, but may also of themselves give rise to structure formation.

4.3.1.   Electrical Fields

One long established effect of applied electrical fields on dispersions is the build up of parallel rows of particle chains in diluted systems. Such behavior has been reported by Muth already in 1927 [126] for milk particles in an alternating electric field, at $2 \cdot 10^4$ to $2 \cdot 10^6$ Hz and a field strength of 333 V $\cdot$ m^{-1}. He explained the formation of such chains of droplets in terms of the polarization of the diffuse electrical double layer. Other examples, and also the theoretical basis of such structures, are described in the book of Efremov [86, 87], referred to previously and also by Mason [127]. In Efremov's book one practical example is given: the occurrence of such structures in the electrophoretic coating process. If branched chains of particles form in the bulk of the electrophoretic bath, the resulting coating becomes gel-like, an undesired state in the coating industry.

In more concentrated systems (above a critical volume fraction, which depends on particle size and on the strength and frequency of the applied field) a continous structure is formed, which, as in all coagulation structures, leads to an increase in viscosity (see section 4.4). Such effects may be monitored using a specially designed, coaxial cylinder viscometer in which the electric field acts perpendicularly to the shear plane; in this way greases have been studied by Vinogradov and Deinega [128].

More recently Ezernack and McLaughlin [129] constructed a two-plate viscometer having the electric field perpendicular to the shear plane. For a dispersion of ethylene-dinitrilotetraacetic acid tetrasodium salt particles in Nujol they measured a twentyfold increase in the viscosity on increasing the voltage from zero to 1.2 kV (gap width $6 \cdot 10^{-4}$ m).

A short summary of the behavior of the so-called "electro-rheological fluids" was given by Stangroom [202].

Some very interesting applications have been reported in the Soviet literature [130], for example, a device for fixing workpieces for machining, consisting of a thin layer (30 μm) of a dispersion of humid silica particles in lubricating oil, which becomes solidlike after application of an electric field and fixes the workpiece [131]. Another application of the so-called "electrorheological effect" is a dielectric motor with liquid moving parts [132].

For a long-range periodic structure Tomita and van de Ven [204] reported changes in the lattice structure (measured from Bragg diffraction) under the action of an external electric field. The electric field acts on the lattice constant in a manner very similar to gravity. In a gravity field, the field acts on the mass of particles, while the electric field acts on the electric charge of the particle.

4.3.2.   Magnetic Fields

Interest in the effects of magnetic fields on particular systems has become increased recently with the invention of the so-called magnetic fluids. Such fluids are ultrastable, ferromagnetic liquids, consisting of single-domain particles, dispersed in a liquid carrier, for example, dispersions of Fe_3O_4, containing particles with a diameter of about 10 nm which are coated with a layer of stabilizer (e.g., oleic acid), which gives rise to a steric layer preventing the coagulation of the particles.

Suspensions of coarse magnetic particles have been in use since the 1940s in magnetic clutches and as indicator fluids for the magnetic testing of materials for defects [133]. One of the first people to prepare a true magnetic fluid was Papell in the early 1960s (as stated by Rosensweig). Rosensweig [134] together with Kaiser prepared magnetic fluids in a number of liquids in the 1960s, which have ten times stronger magnetic properties than Papell's original dispersions. Later in Leningrad in the 1970s Bibik pre-pared such fluids, both on the laboratory scale [136], and also on the semitechnical scale [137]. Now these fluids are prepared all over the world. Displacing the first-used mechanical dispergation by the peptization process reduced the expenses for production by a factor of about 10^4 (!) [138]. In our laboratory, also, magnetic fluids in many different media have been prepared [139—141, 201].

In a magnetic field, a mutual attraction between the particles exists. For the simplest case of two magnetic particles Bibik and Lavrov [142] calculated the magnetic potential energy, in the limiting cases of weak and strong interaction. Scholten and Tjaden [143] extended these calculations to the intermediate range.

Because of the very small size of the particles and the strongly solvated sheath of adsorbed surface active agent, a magnetic fluid, above a critical particle concentration, behaves in a magnetic field like a fluid ferromagnet. Very interesting surface structures are formed under the simultaneous action of gravity, surface tension, and the applied magnetic field [141, 144]. The viscosity of a magnetic fluid increases in the presence of a magnetic field by an amount dependent on the magnitude and direction of the applied field, as was shown theoretically by Hall and Busenberg [145] and experimentally by McTague [146] and by Rosensweig et al. [147]. Rosensweig et al. studied, using a cone-and-plate viscometer having the magnetic field perpendicular to the shear direction, four different magnetic fluids over the following ranges of shear rate $\dot{\gamma}$, magnetic field strength H, and magnetic induction M:

$$0.034 \quad \leqq \dot{\gamma} \leqq 230 \qquad \mathrm{s}^{-1}$$
$$2.87 \cdot 10^4 \leqq H \leqq 1.70 \cdot 10^6 \quad \mathrm{A \cdot m^{-1}}$$
$$5 \cdot 10^{-3} \quad \leqq M \leqq 4.18 \cdot 10^{-2} \, \mathrm{V \cdot s \cdot m^{-2}}$$

Plots were made of the dimensionless groups: $\eta_H/\eta_S = \mathrm{f}(\dot{\gamma} \cdot \eta_{\mathrm{sol}}/M \cdot H)$ (η_H is the viscosity of the magnetic fluid in the presence of the magnetic field, η_S that in its absence, and η_{sol} is the viscosity of the dispersion medium). All the values could be fitted to a single curve, recalculated and redrawn in Figure 4.13. The following conclusions may be drawn:

— On increasing the magnetic energy density ($M \cdot H/2$), the viscosity of the magnetic fluid increases up to nearly 4 times the value without the field present (reaching saturation at very high fields).

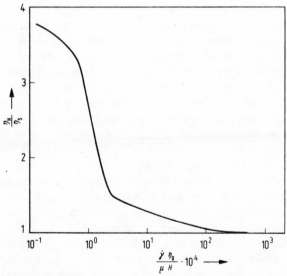

Figure 4.13
Relative increase of viscosity in a magnetic field as function of the dimensionless group $\dfrac{\dot{\gamma} \cdot \eta_0}{M \cdot H}$ (see text) [147].

— The viscosity increase is opposed by an increase in the shear rate $\dot{\gamma}$, indicating that the structure induced by the magnetic field is a coagulation structure, arising from the interparticle magnetic attraction forces.

The effective viscosity, as measured with a capillary viscometer [146], is found to be a function of the direction of the applied field, being about two times greater if the field is parallel to the flow than if it is perpendicular to it.

An effect, also arising from the interaction of magnetic particles in a magnetic field, is the levitation of nonmagnetic objects [134, 138]. In an inhomogeneous magnetic field, nonmagnetic particles are moved in the direction of smaller field strength. It appears in practice as if a magnetic field changes the effective density of a magnetic fluid. By changing the magnetic field strength, nonmagnetic materials of any density may be floated, for example, in the separation of industrial scrap metals, including nonferreous scrap from automobiles or electronic devices. Examples of the many possible applications of magnetic fluids (depending on their structuring in a magnetic field) are given in references [134, 141, 144, 138, 148—150].

4.3.3.   Mechanical Fields (Gravity, Centrifugation, Shear)

Structures in this group are mainly of the repulsive type, that is, periodic structures, in which the mechanical force counteracts the repulsion. Typical examples which are formed under the combined action of gravity and electrostatic repulsion forces are the so-called "Schiller layers" [151—153]. They are formed with platelike particles, such as tungstic acid, and also from rodlike particles, such as the iron oxyhydroxides, as was shown by Maeda and Hachisu [154]. In the latter case, the single particles (rods) are at first oriented parallel to each other, forming "mats", which then grow into the typical Schiller layers, with distances between the horizontal layers of 300—400 nm. The layers form at the bottom of vessels containing the sol, as a result of a very slow coagulation in the lateral direction, sedimentation, and repulsion between the layers. The regular distances between the layers result in iridescence[1]). Periodic structures similar to those discussed in section 4.2.2. may form upon centrifugation of a dispersion, either of charged particles, or of particles with a thick adsorbed layer of solvated macromolecules. Under the action of gravity, the long-range periodic structures, referred to in section 4.2.2., deform only slightly, as reflected by the change of the lattice constants. Because of the very much higher (300—300,000 g) mechanical force in the ultracentrifuge, the distances between the particles are much smaller and no optical effects with visible light are seen. Nevertheless, from consideration of the sediment volume (with and without adsorbed molecules or with different concentrations of electrolyte for charged particles), one may conclude that periodic structures are formed. We measured, for example, with a polystyrene latex with uniform particle size of $D = 480$ nm, with a surface charge density of $1.4 \cdot 10^{-2}$ C $\cdot$ m^{-2}, and an electrokinetic potential of 77 mV at a high electrolyte concentration ($2 \cdot 10^{-2}$ mol $\cdot$ dm^{-3}), a volume fraction of solids in the sediment (at all speeds of the centrifuge) of $\Phi_S = 0.64$, i.e., the value for the random dense packing [59].

---

[1]) The German word for iridescence is "schillern", which has given this type of structure its name.

With the lower electrolyte concentration of $2.35 \cdot 10^{-4}$ mol $\cdot$ dm^{-3} KCl we observed a sediment, the height of which varies with the partial hydrostatic pressure Π (see section 4.4.4.). From equation (4.10), taking $\Phi_{dp} = 0.64$, we may calculate the mean distance between two particle surfaces. The measured values ($\Pi = f(d)$) are numerically in excellent agreement with the theoretical values, calculated by the relation given by Žarkich and Šilov [205]

$$\Pi = \frac{72}{5} \cdot \frac{\Pi_0 \cdot \tilde{\varphi}^2}{\varkappa \cdot a} \cdot \frac{\exp(-\varkappa d)}{1 + \exp(-\varkappa d)} \tag{4.19}$$

$\Pi_0 = 2RTc_0$, is the osmotic pressure of the electrolyte solution (with $c_0 = $ electrolyte concentration in mol $\cdot$ m^{-3}); $\tilde{\varphi}$ is the dimensionless potential.

These experiments are a proof of there being a periodic structure in our centrifuge cell, which is in equilibrium with respect to the external pressure Π and the pressure due to the electrostatic repulsion between the particles.

Table 4.7

Dependence of the distance d between polystyrene particles with adsorbed copolymer of vinyl alcohol and vinyl acetate on the pressure Π

d in nm	Π in Pa
27	3941
27	5707
31	1750
35	1187
38	945
46	667
56	409

If, in another case, we adsorb an organic macromolecule onto the polystyrene particles and measure the partial hydrostatic pressure Π (in the presence of $2 \cdot 10^{-3}$ mol $\cdot$ dm^{-3} KCl, to exclude electrostatic repulsion), we again observe a dependence of Π on d. Assuming a sterically stabilized periodic structure, we may evaluate the compressibility of the adsorbed layer of macromolecule. For example, with an adsorption layer of a copolymer of vinyl acetate and vinyl alcohol we measured the values given in Table 4.7. The steep increase of the pressure at a distance between the particle surfaces of about 25 nm, e.g., a layer thickness of about 12 nm, indicates an almost incompressible layer, which may be taken as the layer of strongly hydrated (about 70% v/v water!) loops.

4.4.   Rheological Behavior of Structured Dispersions

One of the major points of interest about structured dispersions arises from their unusual rheological behavior. From ancient times people have been interested in the kneadability of dough or of clay/water mixtures; in more modern times the curious flow behavior of ketchup (either it stays in the bottle, or it ends up on your trousers) has puzzled people.

Two effects are of specific interest:

1. the reversible, elastic response to small deformations, without any permanent change in structure; and,
2. the shear dependent viscosity as a function of deformation rates, equal to or greater than the rate of rebuilding the structure.

To understand rheological phenomena, it is first necessary to establish the theoretical framework of rheology.

4.4.1.   Types of Rheological Investigations

Rheology is defined as "the study of the deformation of materials, including flow" [155]. The following terms are used to characterize rheological behavior:

1. the external force, which acts tangentially per unit area on the deformed body (= shear stress P in Pa),
2. the deformation γ (see Fig. 4.14),
3. the rate of deformation $\dot{\gamma}$ (= $d\gamma/dt$ in s^{-1}).

Depending upon which terms are held constant, which are varied, and which are measured, various types of rheological experiments may be distinguished:

1. The determination of (apparent) viscosity η (= $P/\dot{\gamma}$ in Pa · s), by measuring the steady state of deformation at constant shear stress (for example, in a capillary viscometer) or by measuring the steady shear stress at constant rate of deformation (in a rotating coaxial cylinder viscometer). In dispersions that are structured by attraction forces between the particles, a steady value results when the velocities of rupture and rebuilding of the structure become equal. In many non-Newtonian systems, the viscosity is a function of the shear stress (or the rate of deformation). The value $\eta' = \eta(\dot{\gamma})$ is therefore called the "apparent viscosity" [155]. For unstructured, Newtonian liquids, η is independent of P or $\dot{\gamma}$.

If the parameter, which was constant, is changed as a function of time [$P = f(t)$ or $\dot{\gamma} = f(t)$], and another parameter recorded, the appropriate flow diagram $P = f(\dot{\gamma})$ or $\dot{\gamma} = f(\beta)$ may be obtained. Often the functions $\eta = f(\dot{\gamma})$ and $\eta = f(P)$ are referred to as "flow diagrams". The shape of the flow diagrams depends on the relationship between the time of measurement and the relaxation time of the structure. If the change of the rate of deformation is faster than that for the formation of an equilibrium structure, one cannot measure any steady value for the viscosity. If the process of structuring lasts longer than the cycle of measurements (from $\dot{\gamma} = 0$ to $\dot{\gamma} = \dot{\gamma}_{end}$) then, on reversing the measurements (from $\dot{\gamma}_{end}$ to $\dot{\gamma} = 0$), a curve results that is different from the initial one.

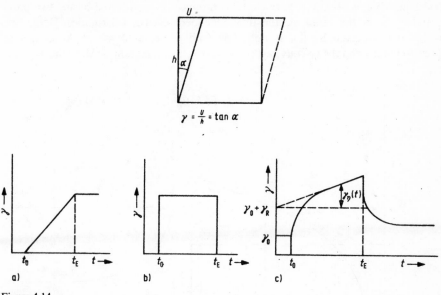

$$\gamma = \frac{U}{h} = \tan \alpha$$

Figure 4.14
Definition of the deformation γ and (a) deformation of an ideal viscous; (b) ideal elastic; and (c) visco-elastic body.

The difference in the areas under both curves (the hysteresis loop) is often used as a measure of the thixotropy of the system.

A comparison of experimental data with theory is only possible if

— the variation of $\dot{\gamma}$ is continuous (or at well-defined time intervals) in both directions;

— the change of $\dot{\gamma}$ in both directions starts from the same point.

The best-defined viscosity results are the equilibrium flow diagrams, constructed by plotting the equilibrium apparent viscosity as a function of P or $\dot{\gamma}$.

2. If P (or $\dot{\gamma}$) at time t_0 is held constant and the variation with time for $\dot{\gamma}$ (or P) is measured, two new methods result: the mesurement of creep deformation or dynamic measurements. If P is very small, so that the structure is not destroyed, the elastic properties of the structure may be determined. Depending on the nature of the structure under study, different $\gamma-t$-curves (Fig. 4.14) result:

— An ideal viscous body. According to Newton's law ($P = \eta \cdot \dot{\gamma}$), at constant P there is a constant rate of deformation (see Fig. 4.14a). All the energy invested in the system during the time interval $t_E - t_0$ is dissipated as thermal energy. At t_E, the state of deformation is not changed.

— For an ideal elastic body, there is a quasispontaneous deformation γ_0, which remains constant until t_E (Fig. 4.14b).

— In practice, most of the suspensions or emulsions show a behavior in between the two ideal types (Fig. 4.14c). The pure elastic component is manifested by the two parts γ_0 and γ_R. γ_0 is the rapid, γ_R the time-dependent elastic deformation [156]. The viscous part is given by the straight line with slope $\tan \delta = 1/\eta$. At t_E γ_0 and γ_R return to zero and the viscous part ($\gamma_\eta = \gamma - \gamma_0 - \gamma_R$) remains at the value reached at t_E.

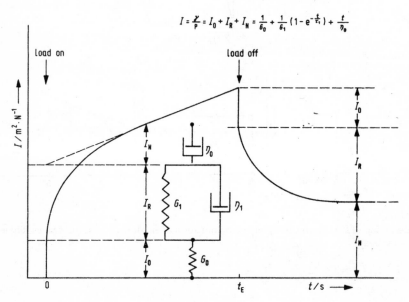

$$I = \frac{\gamma}{P} = I_0 + I_R + I_N = \frac{1}{G_0} + \frac{1}{G_1}(1 - e^{-\frac{t}{\tau_1}}) + \frac{t}{\eta_0}$$

Figure 4.15
Creep diagram (compliance I as function of time t) (see text).

A mathematical description of the deformation—time behavior of such a viscoelastic system may be given in terms of an equation which is useful for mechanical or electrical systems, in which dissipative elements are coupled with nondissipative ones (see Fig. 4.15):

$$\gamma(t) = \frac{P}{G_0} + \frac{P}{G_1} \cdot [1 - \exp(-t/\tau)] + \frac{P \cdot t}{\eta_0} \tag{4.20}$$

This equation contains four parameters, which can be calculated from one $\gamma = f(t)$ measurement: G_0, the modulus of fast elastic deformation (in Pa); G_1, the modulus of the time-dependent elastic deformation (in Pa); τ, the relaxation time for the elastic deformation (in s); and η_0, the creep viscosity (in Pa · s). The apparatus and method for determining the creep deformation is described in reference [157].

We may conclude that there are two fundamental types of methods for the investigation of structured dispersions: investigating the rupture of structure by measuring the viscous behavior, and the study of the undestroyed structure by creep measurements.

4.4.2.   Viscous Behavior of Coagulation Structures

For dilute dispersions of noninteracting solid spheres, the apparent viscosity η' is often described by Einstein's equation:

$$\eta' = \eta_{sol}(1 + k_1 \cdot \Phi) \tag{4.21}$$

(η_{sol} is the viscosity of the solvent). Higher concentrations are taken into account by including terms of higher order of φ, i.e.,

$$\eta' = \eta_{sol}(1 + k_1 \Phi + k_2 \Phi^2 + k_3 \Phi^3 + ...) \tag{4.22}$$

Equations of this type may be reduced to the form, already given by Arrhenius [160]:

$$\eta_r = \eta'/\eta_{sol} = C^x \tag{4.23}$$

(C is a constant, and x is the concentration in % v/v).[1])

Expanding equation (4.23) in a power series, an equation similar to equation (4.22) results, with $k_1 = 100 \cdot \ln C = 4.1$. Einstein found the value for k_1 to be 2.5. A similar approach by Happel [161] led to an equation which gives a somewhat higher constant ($k_1 = 5.5$) for diluted systems and also a somewhat different dependence of η_r on Φ at higher concentrations:

$$\eta_r = 1 + 5.5 \cdot \Phi \cdot \psi \tag{4.24}$$

with

$$\psi = \frac{4\beta^7 + 10 - \dfrac{84}{11}\beta^2}{10(1 - \beta^{10}) - 25\beta^3 \cdot (1 - \beta^4)} \tag{4.25}$$

and $\beta = \Phi^{1/3}$. About 150 different equations, describing the viscous behavior of non interacting dispersions (including these referred to above), were collected, 20 years ago, by Rutgers [162]. As may be deduced therefrom, the viscosity of various dispersions, especially at higher concentrations, cannot be expressed as a single function of the concentration, in spite of the fact that only hydrodynamic effects have to be taken into account. However, noninteracting particles are rarely encountered in practical systems. One can imagine, therefore, how even more difficult a single, all-embracing description becomes, when interparticulate forces are present. If the viscosity (η') is plotted as a function of shear stress, for structured dispersions, at concentrations below that for dense packing, curved plots of the type illustrated in Figure 4.16 results. At very small stresses a high viscosity results, which is independent of the shear stress (Newtonian behavior!); this is called the "creep viscosity" (η_0). At very high stresses, the viscosity again reaches a shear-independent value (η_∞). The behavior in this region may be described by one of the equations for noninteracting particles. Between these two limiting regimes the viscosity is strongly shear dependent (non-Newtonian). In this region one may define two other important parameters: the yield stresses P_{y1} and P_{y2}. The former is associated

---

[1]) $C = 1.042$ for mono- and disaccharides [160].

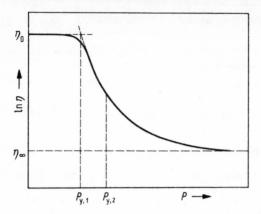

Figure 4.16
General form of a flow curve for a structured dispersion.

with the onset of rupture of the bonds between the particles, the latter by the inflection point of the curve $\eta' = f(P)$ (see Fig. 4.16). P_{y2} is also referred to as the ,,Bingham yield value" and may be evaluated from a $\dot{\gamma} = f(P)$-diagram by extrapolating the linear part at higher shear stresses to $\dot{\gamma} = 0$. The link between the structure of the dispersion and the decrease in viscosity with increasing shear stress had already been observed by W. Ostwald [163], who created the term "Strukturviskosität" (structural viscosity). It was assumed that, with increasing mechanical energy input, more and more of the flocs are broken. This leads to various empirical equations. After some preliminary work by Albers and Overbeek [164], which later was shown to be incorrect [165], a better approach was used by Cross [166, 167] (similar to the older, but empirical work of Oldroyd [168]). He assumed chains of flocculated particles, the length L of which depends on the rate of deformation. At $\dot{\gamma} = 0, L = L_0$ and at $\dot{\gamma} \to \infty, L = 0$. The rupture of bonds is governed by Brownian and shear motion with rate constants (k_0 and $k_1 \cdot \dot{\gamma}^n$, respectively). He further assumes simultaneous rebuilding of the chains at a Brownian rate constant k_2. At equilibrium,

$$dL/dt = k_2 Z - (k_0 + k_1 \cdot \dot{\gamma}^n) \cdot L = 0 \tag{4.26}$$

or

$$L = \frac{k_2 \cdot Z}{k_0 + k_1 \cdot \dot{\gamma}^n} \tag{4.27}$$

and

$$L_0 = \frac{k_2 \cdot Z}{k_0} \tag{4.28}$$

with Z the number of particles per cm³. From equations (4.27) and (4.28),

$$L/L_0 = 1/(1 + \alpha \cdot \dot{\gamma}^n) \tag{4.29}$$

with $\alpha = k_1/k_0$. Hunter [169] proposed that $\alpha = k_1/k_0'$ with $k_0' = k_0 - k_2 \cdot N$ (N is the number of chains per cm^3), based on the fact that the rate of chain formation does not depend on the total number of particles, but only on the number of "free" particles. Equation (4.29) already gives a relationship between the mean aggregate size and the rate of deformation. In addition, one needs a relationship between L and the viscosity of the dispersion. By analogy with the viscosity of solutions of long-chain polymers [170], one may write

$$\eta = \eta_\infty + BL \tag{4.30}$$

and

$$\eta_0 = \eta_\infty + BL_0 \tag{4.31}$$

From (4.30) and (4.31)

$$\frac{L}{L_0} = \frac{\eta - \eta_\infty}{\eta_0 - \eta_\infty} \tag{4.32}$$

or with (4.29)

$$\eta = \eta_\infty + \frac{\eta_0 - \eta_\infty}{1 + \alpha \gamma^n} \tag{4.33}$$

or rearranging

$$\frac{\eta_0 - \eta_\infty}{\eta - \eta_\infty} \quad 1 = \alpha \cdot \gamma^n \tag{4.34}$$

Plots of $\log\left[\dfrac{\eta_0 - \eta_\infty}{\eta - \eta_\infty} - 1\right]$ as a function of $\log \gamma$ should be linear, with slope n and $\log \alpha$ as the intercept on ordinate axis. The constant α, which contains the rate constants for rupture and rebuilding, is assumed to be a measure of the inter-particle attraction. By way of example, the flow curve data for a polyvinyl acetate latex (see Fig. 4.17) are analyzed in terms of equation (4.34). In the range $100\ \text{s}^{-1} \geqq \gamma \geqq 0.1\ \text{s}^{-1}$ a straight line results with slope 0.7 (Cross [167] had shown that in most cases n is of the order of 2/3, i.e., abaut 0.7). α depends strongly on the exact choice of η_0 and η_∞. Reliable estimates are required to obtain accurate values for α. In the example given above, as in most reported flow diagrams, there is some doubt about the values of η_0 and η_∞ (to obtain a complete flow diagram, as in Figure 4.17, one has to make measurements on the sample in various devices). Equations (4.33) and (4.34) therefore can only be regarded as being semiempirical.

A better approach was undertaken by Michaels and Bolger [171]. These authors divided the destruction of the structure into different steps: deformation of the network (shear stress P_N), rupture of bonds (P_R), and viscous (hydrodynamic interaction (P_V). The total rate of energy dissipation per unit volume is

$$E_{tot} = E_N + E_R + E_V \qquad (\text{in } J \cdot m^{-3} \cdot s^{-2}) \tag{4.35}$$

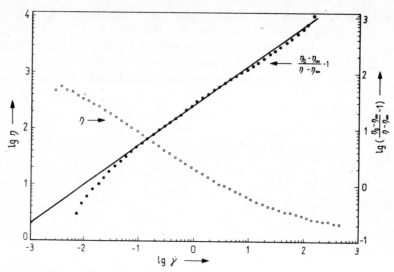

Figure 4.17
Flow diagram and plot after Cross [166] for a polyvinyl acetate latex.

$(E_i = P_i \cdot \dot{\gamma})$. Two limiting equations were obtained for low and high rates of deformation, respectively:

$$E_{tot} = \frac{A \cdot \dot{\gamma} \cdot \Phi_F^2}{2\pi^2 D_F^2 d_0} + \eta_{sol} \cdot \dot{\gamma}^2 (1 + 2.5\Phi_F) \quad \text{(high)} \tag{4.36}$$

$$E_{tot} = \frac{F(\dot{\gamma}) \cdot A \cdot (\Phi_F - \Phi_{F0})^3}{24 d_0 D_F}$$

$$+ \frac{5 \cdot \dot{\gamma} \cdot A C_{AF}^2 \cdot \Phi_F^2}{8\pi \cdot D_A \cdot D_F \cdot d_0 \cdot (C_{AF} - 1)} + \eta_{sol} \cdot \dot{\gamma}^2 (1 + 2.5\Phi_A) \quad \text{(low)} \tag{4.37}$$

where Φ_F is the volume fraction of primary particles (not destroyed at the highest possible rates of deformation); Φ_{F0} minimal volume fraction of primary particles, necessary for structure formation (identical with Φ_{crit}); Φ_A volume fraction of the aggregates; $F(\dot{\gamma})$ a dimensionless form factor, which depends on the nature of the structure (not clearly defined); A Hamaker constant for the attraction energy between particles; D_F, D_A diameter of primary particles, of aggregates, respectively, d_0 equilibrium distance between the particle surfaces; $C_{AF} = \Phi_A / \Phi_F$, i.e., volume fraction of solids in the single aggregates.

At high rates of deformation one obtains a straight line for $P = \mathrm{f}(\dot{\gamma})$, because A, Φ_F, D_F, and η_{sol} are independent of $\dot{\gamma}$. The slope of this line is equal to the plastic viscosity

$$\eta = \eta_{sol}(1 + 2.5\Phi_F) \tag{4.38}$$

The ordinate at $\dot{\gamma} = 0$ is the Bingham yield value

$$P_{y2} = P_B = \frac{A \cdot \Phi_F^2}{2 \cdot \pi^2 \cdot D_F^2 \cdot d_0} \tag{4.39}$$

Equation (4.37) is difficult to check at low $\dot{\gamma}$, because $F(\dot{\gamma})$ is not known, and it is necessary to measure Φ_A and D_A for every value of $\dot{\gamma}$; this has only been done for a few systems [169, 170].

Firth and Hunter [172] developed this model further by taking into acount the elastic deformation of the flocs in flow ("elastic floc model"). They showed from many experiments the possibilities and the difficulties involved in interpreting experimental data using equations (4.35)—(4.39). One advantage of equation (4.39) is the possibility of expressing the readily determined Bingham yield stress in terms of interaction energies. With known values for Φ_F and D_F one obtains A/d_0. For many materials the Hamaker constant A is known [26, 173] and one may then calculate the equilibrium distance between particles. For example, Michaels and Bolger [171], assuming $d_0 = 0.5$ nm, obtained a value for the Hamaker constant of kaoline in water of 10^{-17} to 10^{-18} J, which is about three orders of magnitude higher than expected. This result is not very surprising since they extrapolated from a regime where the structure is fully broken, to the situation where the rupture begins. In addition, they ignored the electrostatic interactions. Firth and Hunter introduced the electrostatic interaction into equation (4.39). However, they have not been able to derive an analytical equation for the Bingham yield value as a function of the complete interaction energy.

The measurement of the flow curves $[\dot{\gamma} = f(P)$ or $\eta = f(P)]$ is possible with standard methods, such as Couette-type (coaxial cylinder) viscometers, or capillary viscometers with variable pressure. However, for the extreme parts of the flow diagrams (η_0 and η_∞), it is in most cases necessary to use special methods (creep and high-shear measurements).

The yield value of plastic materials may also be measured using a simple method: the conical plastometer. This method, which has been widely used, in various forms, in scientific and industrial laboratories in the Soviet Union, is less known in the West. Hence a fairly detailed description of this method is given here.

The conical plastometer, introduced by Rehbinder and Semenenko [174], is a variation of the more widely used penetrometer. It is based on the "Kegelprobe" introduced by Ludwik [175], for testing the hardness of metals. One uses a conical probe, which is placed with a given constant load F, on the surface of the sample. If a yield stress exists, the cone penetrates into the sample to a depth h, at which the shear stress at M, i.e., the surface area of the cone in contact with the sample is equal to the yield value P_y:

$$F \cdot \cos \alpha = P_y \cdot M \tag{4.40}$$

with $\alpha = $ (cone tip angle)/2.

From simple geometry one obtains for the area of contact between the cone and the sample, at depth h,

$$M = \frac{\pi \cdot h^2 \cdot \tan \alpha}{\sin \alpha} \tag{4.41}$$

The yield stress follows from equations (4.40) and (4.41) together with the standard equation $F = m \cdot g$;

$$P_y = \frac{m}{h^2} \frac{g \cos^2 \alpha \cdot \cot \alpha}{\pi} \tag{4.42}$$

Setting

$$K = \frac{g \cos^2 \alpha \cdot \cot \alpha}{\pi} \tag{4.43}$$

one obtains

$$P_y = K \frac{m}{h^2} \tag{4.44}$$

For the most used cone angle with $2\alpha = 60°$, the constant becomes $K = 4.06 \text{ m} \cdot \text{s}^{-2}$.

Deviations from equation (4.44) are possible if a layer of "liquified" structure appears, which changes the effective size of the cone. This can be avoided by a very low velocity of penetration of the cone. If the part of the dispersion that is displaced by the cone forms a wall around the cone, the real penetration depth is difficult to obtain, i.e., an exact measurement becomes impossible [176].

The best way to estimate P_y from the depth of penetration is to plot m as a function of h^2 for at least 4—5 points. This will show whether equation (4.44) is applicable, since a straight line through the origin should result. Furthermore, one can omit doubtful experimental points, and, using the method of least squares to fit the data, one minimizes the error in the calculated yield stress.

In the practical construction of the apparatus, the cone may be suspended to compensate for its weight, the penetration depth can be measured by an optical device or by an inductive transducer, and the load can be increased automatically in suitable steps [177] (see Fig. 4.18).

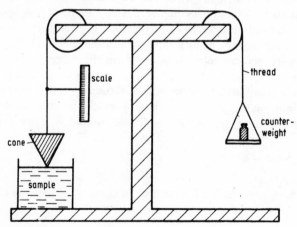

Figure 4.18
Simple forms of a conical plastometer.

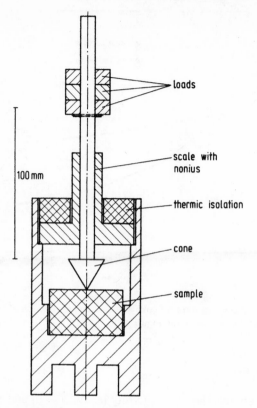

loads

scale with nonius

thermic isolation

100 mm

cone

sample

Figure 4.19
Specially designed simple conical plastometer for coke—tar mixtures.

For experiments with more solidlike materials, at elevated temperatures (for example, the "green masses" in the production of graphite electrodes) the authors have developed a simpler device, machined from brass (for good conduction of heat from an oil bath) [178] in which the load is placed directly on the cone shaft and the depth of penetration is measured at the shaft with the aid of a vernier dial (see Fig. 4.19).

Using this simple apparatus, among others the yield stress of a mixture of coke and coal tar was determined as a function of the temperature (see Fig. 4.20).

4.4.3.   Elastic Behavior of Coagulation Structures

To ask for the initial nature of the structure after its total destruction is akin to fortune telling. A much better approach is to look at the response of the undestroyed structure to mechanical stresses. One way is to carry out creep experiments, i.e., measurements of the deformation-time response to a constant small stress. Another possibility would be the compression of the structured systems by direct external pressure or by centrifugation.

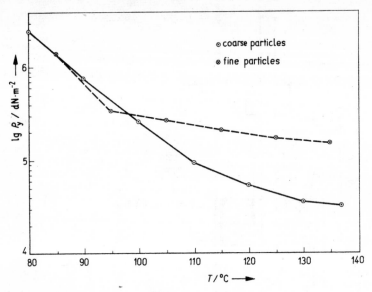

Figure 4.20
Yield stress of "green mass" (mixtures of coke and coal tar) as function of temperature, measured with the plastometer shown in Figure 4.19.

4.4.3.1.   Creep Measurements

From creep experiments one obtains the creep viscosity (η_0 for $\dot{\gamma} \to 0$) and the elastic response to small stresses [see equ. (4.20)]. The creep viscosity, which is typically many orders of magnitude greater than the viscosity of the deaggregated system η_∞ is in dependent of the shear stress, implying apparently a Newtonian viscosity. Nevertheless, it is not Newtonian in strict rheological terms (i.e., resulting from the internal friction between molecules and particles), but results from the rupture and reforming of bonds, as well as "gliding" (a sort of particle surface diffusion) of the particles into new positions, resulting in reorientation of the network.

Elastic deformations, as mentioned above (section 4.4.1.), are reversible. One may therefore use a thermodynamic approach to explain the relation between shear stress and changes in structure.

The use of creep experiments in colloid chemistry was intensively influenced by Rehbinder [179], based on the first experiments carried out by Švedov [180]. The following conditions are essential for the measurement of creep deformation:

— The dispersion has to be studied in a geometrically well-defined space with an accurately measured, small shear stress.
— The resulting, very small deformation also has to be measured with great accuracy as a function of time.

Both types of axially symmetrical viscometers, i.e., the coaxial-cylinder and the cone-and-plate type may be used. One of the walls (the outer cylinder or the plate) is fixed.

A constant torque is applied along the axis of the moving part, without disturbing the axial orientation of the walls with respect to each other. The very small rotation of the movable part has to be measured and recorded. The authors have used a specially designed cone-and-plate type creep viscometer [157], for which the following equations hold (see Fig. 4.21).

The shear stress at the cone is given by

$$P = \frac{M}{2\pi R^3} \qquad (4.45)$$

The torque,

$$M = mg\,\frac{D}{2} \qquad (4.46)$$

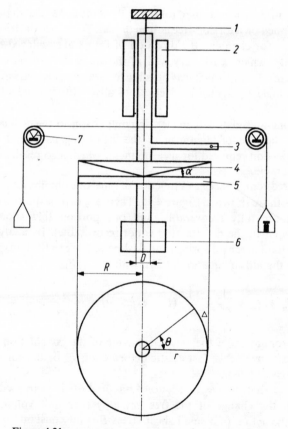

Figure 4.21
Schematic view of the cone-and-plate creep viscosimeter:
1, torsion-free thread; 2, air bearing; 3, holder for iron core of the inductive transducer; 4, cone; 5, plate; 6, motor for rotating the plate before measurements; 7, wedge bearings.

Hence,

$$P = \frac{mgD}{4\pi R^3} \tag{4.47}$$

The strain is given by

$$\gamma = \frac{\tan \Theta}{\tan \alpha} \tag{4.48}$$

and the compliance by

$$\frac{\gamma}{P} = I = \frac{\Theta 4\pi R^3}{m\alpha g D} = \frac{\Theta}{m} K \tag{4.49}$$

An almost frictionless air bearing was used for the shaft of the cone. The shaft, together with the cone, was suspended on a torsion-free cotton thread. To reduce the friction in the system further, the rollers for changing the direction of forces are replaced by wedge-bearings 7 (see Fig. 4.21), which work very well at the small deflections of the wall (Δ maximal reaches 0.1 cm). In this way it was possible to measure shear stresses as small as $0.003 \text{ N} \cdot \text{m}^{-2}$. The angle Θ is measured electronically, with the aid of a carrier frequency bridge.

The benefits of the cone-and-plate viscometer, in comparison to the coaxial cylinder viscometer, are, firstly, the ease of alignment of the measuring parts and, secondly, the small volume of dispersion required for measurement, which also results in a better thermal control of the sample.

If one plots the measured compliance I as a function of time, one obtains for coagulation structures results such as those shown in Figure 4.15. There are two elastic portions of the curve (I_0 and I_R) that are fully recoverable, and one portion that shows viscous behavior (energy is dissipated, the deformation is not recoverable). By analysis of this behavior with the much used model of springs and dashpots, one can fit the experimental values for $I = f(t)$ with the aid of equation (4.20) (with $I = \gamma/P$):

$$I = I_0 + I_R + I_N = \frac{1}{G_0} + \frac{1}{G_1} (1 - \exp(-t/\tau)) + \frac{t}{\eta_0} \tag{4.20a}$$

For the theoretical interpretation of the elastic behavior of the coagulation structure, only the time-independent, reversible, fast elastic deformation γ_0, or the corresponding modulus ($G_0 = P/\gamma_0$) is considered.

If a structure, such as that shown in Figure 4.6a or 4.6b is deformed (without any rupture of bonds), the change of the free enthalpy (per unit volume), at the deformation γ_0, under the action of a small shear stress P is (at constant composition, volume, temperature, and pressure given by [181]

$$\left(\frac{\partial G_0}{\partial \gamma}\right)_{\mu, V, T, p} = \left(\frac{\partial U}{\partial \gamma}\right)_{\mu, V, T, p} - T \left(\frac{\partial S}{\partial \gamma}\right)_{\mu, V, T, p} = P \tag{4.50}$$

with the internal energy equal to the interaction energy

$$U = V_a \tag{4.51}$$

The total entropic contribution is associated with two effects: the change in conformational entropy of the structure and the change in entropy of a layer of solvent (or of an adsorbed layer) between the particles, as a consequence of a change in their separation. This second contribution has not been considered much, until recently, owing mainly to the lack of knowledge of the structure of solvent layers.

For the conformational entropy Šukin and Rehbinder [182] used

$$-T \left(\frac{\partial S}{\partial \gamma} \right)_{\mu, V, T, p} = P = 3nk_B T \tag{4.52}$$

Hence, one obtains for the modulus G_0,

$$G_0 = \frac{P}{\gamma_0} = 3nk_B T \tag{4.53}$$

If all the particles contribute to the entropy change, one obtains, using equation (4.53), for a dispersion of Aerosil 200, 4.04% v/v, a theoretical value for $G_0 = 551$ N · m^{-2} (with $n = 6 \cdot \Phi/\pi \cdot D^3$, $D = 12 \cdot 10^{-9}$ m, $k_B = 1.38 \cdot 10^{-23}$ J · K^{-1}, and $T = 298$ K). Experimentally for this particular system in decane, $G_{0, exp}$ was found to be $1.2 \cdot 10^4$ N · m^{-2}, and in water 373 N · m^{-2} [80].

As may be seen from these experiments, the entropic contribution alone cannot explain the values measured in decan. In water, which is a much more structurable system, an entropic contribution seems more probable. If the entropic term is neglected, then

$$\left(\frac{\partial G_0}{\partial \gamma} \right) = \left(\frac{\partial U}{\partial \gamma} \right) = P \tag{4.54}$$

or in integrated form

$$V_a = U = P \cdot \gamma_0/2 \tag{4.55}$$

V_a is the change of energy, per unit volume, at deformation γ_0.

It was assumed [183] that at shear deformation γ_0 for a unit cube of the network (Fig. 4.14) the diagonal increases by $\frac{\sqrt{2}}{2} \cdot \gamma_0$. If the undeformed diagonal is considered to be a chain of particles, there are $\sqrt{2}/(D + d_0)$ particles in the diagonal chain (d_0 the equilibrium distance between the particle surfaces). The particles are not deformable; therefore an increase in chain length is only possible by stretching the bonds between the spheres (increasing d). In this experiment the particle chains are orientated in one direction by rotation of the plate, before the static elastic measurements are performed.

$$\Delta d = \frac{\text{increase of diagonal length}}{\text{number of bonds per diagonal}}$$

$$\Delta d = \frac{\gamma_0(D + d_0)}{2} = \frac{P}{G} \cdot \frac{(D + d_0)}{2} \tag{4.56}$$

The interaction energy, $V_a = P \cdot \gamma_0/2$, acts on n bonds per unit volume, with

$$n = \frac{6\Phi}{\pi D^3} \tag{4.57}$$

Therefore one obtains for the energy change per bond, at deformation γ_0

$$\Delta V_b = \frac{P\gamma_0\pi D^3}{12\Phi} \tag{4.58}$$

For the modulus, we obtain from (4.56) and (4.58)

$$G_0 = \frac{6\Phi}{\pi D^3} \cdot \frac{(D + d_0)\,\Delta V_b}{\gamma_0\,\Delta d} \tag{4.59}$$

or, if $D \gg d_0$ (reasonable for particles in the primary minimum),

$$G_0 = \frac{6\Phi\,\Delta V_b}{\pi D^2\,\Delta d\,\gamma_0} \tag{4.60}$$

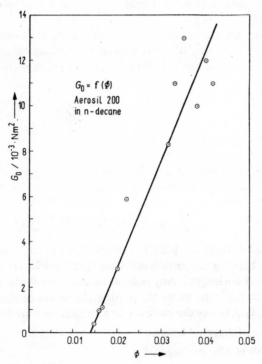

Figure 4.22

Moduli G_0 for dispersions of Aerosil 200 in n-decane as a function of the volume fraction of solids Φ.

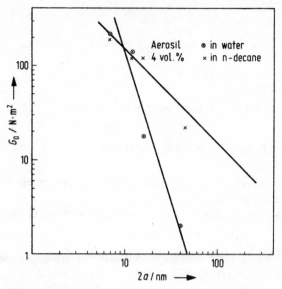

Figure 4.23
Dependence of the modulus G_0 on the diameter of silica particles $2a$, measured in n-decane and in water.

In decane, only dispersion forces operate. These may be calculated approximately using

$$\frac{\Delta V_D}{\Delta d} = \frac{A^* \cdot D}{24 \cdot d_0^2} \tag{4.61}$$

which gives for G_0

$$G_0 = \frac{\Phi A^*(D + d_0)}{4\pi D^2 \gamma_0 d_0^2} \tag{4.62}$$

or for $D \gg d_0$

$$G_0 = \frac{\Phi A^*}{8 \cdot \pi D d_0^2 \cdot \gamma_0} \tag{4.63}$$

In order to test the validity of equation (4.63), the dependence of G_0 on Φ (Fig. 4.22) and on $1/D$ (Fig. 4.23) was measured [31, 39, 80], for Aerosil in decane. As one can see from the figures, both the linear dependence of G_0 on Φ and its inverse proportionality to D (the slope of the straight line in Fig. 4.24 is -1) are obtained.

From the slopes of the functions $G_0 = f(\Phi)$ and $G_0 = f(1/D)$,

$$\frac{dG_0}{d\Phi} = \frac{1}{8\pi D\gamma_0}\frac{A^*}{d_0^2} = 4.46 \cdot 10^5 \text{ N} \cdot \text{m}^{-2} \tag{4.64}$$

and

$$\frac{dG_0}{d(1/D)} = \frac{\Phi A^*}{8\pi\gamma_0 d_0^2} = 1.35 \cdot 10^{-4} \text{ J m}^{-2}$$

or for

$$\frac{A^*}{d_0^2} = 0.13\gamma_0 \text{ J} \cdot \text{m}^{-2} \quad \text{(from } G_0 = f(\Phi)) \tag{4.65}$$

$$\frac{A^*}{d_0^2} = 0.08\gamma_0 \text{ J} \cdot \text{m}^{-2} \quad \text{(from } G_0 = f(1/D))$$

The good agreement between the two values, obtained from independent sets of measurements, justifies the use of equation (4.63).

Furthermore, from the mean value of $A/d_0^2 = 0.11\gamma_0$ J $\cdot$ m^{-2}, using the relationship

$$A^* = (A_1^{1/2} - A_0^{1/2})^2 = 1.5 \cdot 10^{-21} \text{ J} \tag{4.66}$$

with A_0 the Hamaker constant for decane $= 5.0 \cdot 10^{-20}$ J [185]; A_1 the Hamaker constant for silica $= 3.4 \cdot 10^{-20}$ J [173]; and $\gamma_0 = 10^{-2}$, one obtains $d_0 = 1.36 \cdot 10^{-9}$ m, which seems to be a reasonable value. Another test of the validity of equation (4.63) is to measure G_0 for dispersions in decane ($A = 1.5 \cdot 10^{-21}$ J) and in a 3 mol $\cdot$ dm^{-3} aqueous solution of potassium chloride ($A = 6.2 \cdot 10^{-21}$ J, the same as for water) where electrostatic repulsion and structure effects are absent. If only dispersion forces are acting, the ratio

$$\frac{A^*}{G_0} = \frac{8\pi D \cdot d_0 \cdot \gamma_0}{\Phi} \tag{4.67}$$

should be constant. The respective values for dispersions in decane, 3 mol $\cdot$ dm^{-3} KCl solution, and in water are shown in Table 4.8. As support for the conjecture that the energetic part of the interaction in decane and other hydrocarbons is the dominant contribution, measurements of the temperature dependence of the elasticity of structured dispersions of Aerosil 200 in decane were made [184][1]. The value for the change of the modulus with temperature was found to be $\Delta G_0/\Delta T$ from 150 to

Table 4.8
Elastic modulus and A^*/G_0 (see text) for different dispersions

Dispersing medium	G_0 in (N $\cdot$ m^{-2})	A^* in J	A^*/G_0 in m^3
n-Decane	$1.2 \cdot 10^4$	$1.5 \cdot 10^{-21}$	$1.25 \cdot 10^{-25}$
3 mol $\cdot$ dm^{-3} KCl solution	$5.5 \cdot 10^3$	$6.2 \cdot 10^{-22}$	$1.12 \cdot 10^{-25}$
Water	$3.7 \cdot 10^2$	$6.2 \cdot 10^{-22}$	$1.67 \cdot 10^{-24}$

[1]) In this paper, the value for G_0/T is not correct, owing to an error in the calculation of density [184a].

$170 \text{ N} \cdot \text{m}^{-2} \cdot \text{K}^{-1}$. If one only considers the entropic contribution, i.e.

$$\frac{\Delta G_0}{\Delta T} = 3 \cdot n \cdot k_\text{B}$$

one calculates a theoretical value of only $0.92 \text{ N} \cdot \text{m}^{-2} \cdot \text{K}^{-1}$. The strong dependence of the elastic modulus on temperature was, therefore, interpreted as being due mainly to the increase of the van der Waals-Hamaker interaction with increasing temperature. "

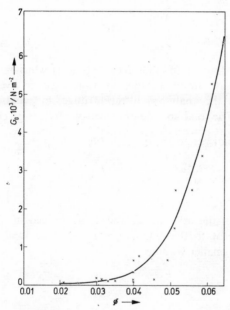

Figure 4.24
Moduli G_0 for dispersions of Aerosil 200 in water as a function of the volume fraction of solids Φ.

In water, a nonlinear dependence of G_0 on Φ is found (see Fig. 4.24), because, as mentioned above, the entropic contribution must also taken into account. For unknown systems it is therefore useful to use the complete expression for the elastic modulus:

$$G_0 = n \left(3k_\text{B}T + \frac{D + d_0}{\gamma_0} \frac{\Delta V}{\Delta d} \right) \tag{4.68}$$

The term $\Delta V / \Delta d$ includes not only the dispersion forces, but also in general any electrostatic repulsion or steric forces.

These experiments support the hypothesis of a network structure, built up from particle chains.

4.4.3.2.   Compression of Structured Dispersions

If a coagulation structure is subjected to a centrifugal field, a pressure is created which depends on the distance from the axis of rotation r, the angular velocity ω, and the mass concentration $c = \Phi \, \Delta\varrho$.

By analogy with the discussion of the ,,hydrostatic partial pressure" in gels, given by Svedberg [186], one may write

$$\Pi = \omega^2 \int_{r_i}^{r_n} \Phi_s \, \Delta\varrho \, r \, dr \tag{4.69}$$

$$\Pi = \frac{\omega^2}{2} \, \Phi_s \, \Delta\varrho \cdot (r_n^2 - r_i^2) \tag{4.70}$$

with $\Delta\varrho = \varrho_p - \varrho_0$. Φ_s (the volume fraction of solids in the sediment), which is, in practice, a function of r_i. An approximate mean volume fraction may be evaluated for a sector-shaped cell (as used in the analytical ultracentrifuge, to avoid convection flows) in the following manner: The total volume of dispersed phase is given by

$$v_{tot} = \text{(volume of the dispersion in the cell)} \cdot \Phi_0$$

or

$$v_{tot} = \Phi_0 b h_0 \frac{a_0 + a_n}{2} \tag{4.71}$$

with $h_0 = r_n - r_0$; $b =$ depth of the cell; a_0, a_i, and a_n as defined in Figure 4.25. After compression of the sediment to the height $h_i = r_n - r_i$, the total volume of particles is now contained in the smaller volume

$$v_i = b \cdot h_i \cdot \frac{a_i + a_n}{2} \tag{4.72}$$

The volume fraction of solids in the compressed sediment therefore becomes

$$\Phi_s = \frac{v_{tot}}{v_i} = \frac{h_0}{h_i} \, \Phi_0 \, \frac{a_0 + a_n}{a_i + a_n} \tag{4.73}$$

$a = r \tan \alpha$ (see Fig. 4.25); hence,

$$\frac{a_0 + a_n}{a_i + a_n} = \frac{r_0 + r_n}{r_i + r_n} \tag{4.74}$$

and

$$\Phi_s = \frac{h_0}{h_i} \, \Phi_0 \, \frac{r_0 + r_n}{r_i + r_n} = \Phi_0 \, \frac{r_n^2 - r_0^2}{r_n^2 - r_i^2} \tag{4.75}$$

With equations (4.70) and (4.75) one obtains

$$\Pi = \frac{\omega^2}{2} \, \Delta\varrho\Phi_0(r_n^2 - r_0^2) \tag{4.76}$$

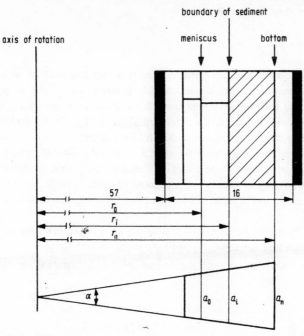

Figure 4.25
Definition of distances in the analytical ultracentrifuge.

Buscall [187] gives an analytical expression for the relation between the above obtained pressure Π and the volume fraction of solids in the compressed sediment [Φ_s, calculated from the centrifugal data, using equation (4.75)], which results in the definition for a "compression modulus" for the structure. The strain on compression from a volume v_1 to a smaller volume v_2 is given by

$$\text{strain} = -\int_{v_1}^{v_2} \frac{\mathrm{d}v}{v} = \ln \frac{v_1}{v_2} \tag{4.77}$$

For a consolidating sediment one may write, by analogy for a compressible material, having a bulk modulus

$$K(v) = -\frac{\mathrm{d}\Pi}{\mathrm{d}\ln v} \tag{4.78}$$

$K(v)$ is an irreversible function of the occupied volume, since no spontaneous expansion is observed after deloading. The modulus therefore is not strictly a compression modulus as for compressible materials. Nevertheless, this term (or better, network modulus) may be used for the compression of the coagulation structure.

As $v \approx 1/\Phi$, one may also write

$$K(\Phi) = \frac{d\Pi}{d \ln \Phi_S} \qquad (4.79)$$

If the difference in concentrations between the top and the bottom of the sediment is not large, one may use the mean volume fraction of solids in the sediment [equation (4.75)]. The authors have measured the sediment heights as a function of the angular velocity of the rotor in an ultracentrifuge [188], up to 60,000 rpm, i.e., $2.6 \cdot 10^5$ g. From the height at zero rpm h_0 and at compression h_i one may calculate Φ_S and obtain the modulus $K(\Phi)$ from the plot $P = f(\ln \Phi)$. Buscall has shown that the network modulus of structured dispersions is related to the elastic modulus, determined rheologically. As example, results are presented of some experiments on a dispersion of Aerosil 200 in decane, and in water, which have been investigated with the creep method also.

From Figure 4.26 it can be seen qualitatively that similar behavior results for the $K(\Phi) = f(\Phi)$ diagram and for the $G_0 = f(\Phi)$ diagram, respectively:

— All K and G_0 values increase with increasing solid concentration, in decane as well as in water.

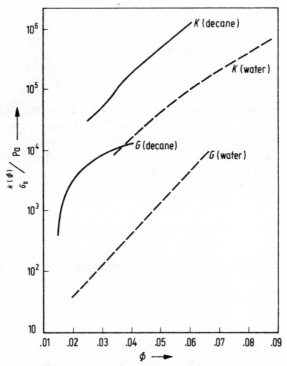

Figure 4.26
Comparison of network moduli $K(\Phi)$ with shear moduli G_0 for dispersions of fumed silica in decane and in water as a function of the volume fraction of solids Φ.

— The values for K and G_0 in decane are, for a given Φ, greater than for the dispersions in water.

— The network moduli always exceed the shear moduli.

These results may be interpreted only phenomenologically. As had been shown before (section 4.4.3.1.), the shear modulus depends on the number and strength of bonds between the particles. The number of bonds depends, at constant particle size, linearly on the solid concentration, i.e., G_0 should increase with increasing Φ (as found experimentally). The strength of the single bond is greater in decane than in water, owing to the greater Hamaker constant for SiO_2 in decane in respect to the value in water. Hence, the shear modulus should be greater in decane than in water. This also was verified experimentally. In contrast with the reversible process of shear deformation, the compression is irreversible, an essential part of energy is dissipated for the re-formation of destroyed bonds. The network modulus should, therefore, be greater than the shear modulus, as had been measured.

Buscall [187] some years ago published results, showing the equality of $K(\Phi)$ and G, that seem to contradict our measurements.

In a recent paper by Steward and Sutton [206][1]) different values for G and K have been published (orders of magnitude higher), the network modulus being about one order of magnitude greater than the shear modulus,[2]) in accord with our measurements.

Compression experiments on Aerosil dispersions in aliphatic alcohols have also shown a close relationship with the mechanical strength of the structure [189].

The compression of periodic (repulsive) structures has also been studied, using centrifugation [190] (see section 4.3.3.), and also by the method, developed by Barcley et al. [191], in which a suspension of polystyrene latex particles is confined between a semipermeable filter and an impermeable plastic membrane. An externally applied pressure forces the dispersing medium through the filter. The volume fraction is measured as a function of the applied pressure. Hence the distance between the equidistant particles may be calculated. One may then construct a repulsive force-distance diagram. The experimental technique was improved by Homola [192] and later modified by Dickinson [193], who studied repulsive forces due to adsorbed layers as well as those due to electrostatical repulsion [194]. In the latter case the experimental results were compared with theoretical models, and good agreement was obtained. For the attraction energy the following equation was used:

$$V_D = - \frac{A}{6} \left[\frac{2a^2}{d^2 + 4ad} + \frac{2a^2}{d^2 + 4ad + 4a^2} + \ln \frac{d^2 + 4ad}{d^2 + 4ad + 4a^2} \right] \qquad (4.80)$$

For the electrostatic repulsion, it was found to be necessary to use the following relationship:

$$V_R = 2\pi\varepsilon\varepsilon_0 a\psi_0^2 \ln \left(1 + \exp \left(-\varkappa d \right) \right) \qquad (4.81)$$

---

[1]) From the same laboratory as Buscall, on the same objects (520-nm polystyrene latex, flocculated with 0.1 mol · dm^{-3} electrolyte), and referring to [187].

[2]) As given in Figs. 3 and 4 in ref. [206], in the text again the equality of K and G is discussed!

rather than the alternative form,

$$V_{\mathrm{R}} = \frac{32\pi\varepsilon\varepsilon_0 k_{\mathrm{B}}^2 T^2 \gamma^2 a}{e^2 z^2} \exp(-\varkappa d) \tag{4.82}$$

with

$$\gamma = \tanh \frac{ze\psi_0}{4k_{\mathrm{B}}T}$$

It seems that equation (4.81) is a good approximation for concentrated systems.

In an independent paper by Feldkamp et al. [195] a similar type of cell was used for the pressure range of 200 to 1800 Pa in a study of aluminium hydroxy-carbonate gels. These authors showed that it is possible to distinguish between different types of structures (periodic or coagulation) from pressure-volume fraction plots.

References

Chapter 1

[1] Loeb, A. L., P. H. Wiersema, and J. Th. G. Overbeek, Double Layer around a Spherical Colloid Particle, MIT Press, Cambridge, Massachusetts 1961

[2] Debye, P., and E. Hückel, Phys. Z. **24**, 185 (1923)

[3] Martynov, G. A., Kolloidny Ž. **38**, 1111 (1976)

[4] Duchin, S. S., N. M. Semenichin, and Z. M. Žapinskaja, Dokl. Akad. Nauk SSSR **193**, 385 (1970)

[5] Sigal, V. L., S. S. Duchin, and V. E. Samanskij, Dokl. U. Akad. Nauk SSSR **4**, 346 (1970)

[6] Sigal, V. L., S. S. Duchin, and V. E. Samanskij, Poverchnostnye sily v tonkich plenkach, Moscow 1971, p. 79

[7] Duchin, S. S., and B. V. Derjaguin, in Surf. Colloid Sci., ed. by E. Matijvic **7** (1974)

[8] Derjaguin, B. V., Acta Physicochim. URSS **10**, 333 (1939)

[9] Derjaguin, B. V., Kolloidny Ž. **6**, 291 (1940)

[10] Derjaguin, B. V., Kolloidny Ž. **7**, 285 (1941)

[11] Derjaguin, B. V., and L. D. Landau, Acta Physicochim. URSS **14**, 633 (1941)

[12] Verwey, E. J., and J. Th. G. Overbeek, Theory of the Stability of lyophobic Colloids — Elsevier, New York, Amsterdam 1948

[13] Derjaguin, B. V., Kolloid Z. **69**, 155 (1939)

[14] McCartney, L. N., and S. Levine, J. Colloid Interface Sci. **30**, 345 (1969)

[15] Ohshima, H., D. C. Chan, T. W. Healy, and L. R. White, J. Colloid Interface Sci. **92**, 232 (1983)

[16] Chan, B. K., and D. C. Chan, J. Colloid Interface Sci. **92**, 283 (1983)

[17] Hogg, R., T. W. Healy, and D. W. Fuerstenau, Trans. Faraday Soc. **62**, 1638 (1966)

[18] Hamaker, H. C., Rec. Trav. Chim. **55**, 1015 (1936)

[19] Hamaker, H. C., Rec. Trav. Chim. **56**, 727 (1937)

[20] Hamaker, H. C., Physica **4**, 787 (1937)

[21] Lifschitz, E. M., Sov. Phys. JETP **29**, 94 (1955)

[22] Dcialožinskij, J. E., E. M. Lifschitz, and L. P. Pitaevski, Sov. Phys. JETP **36**, 1797 (1959); **37**, 229 (1959)

[23] Overbeek, J. Th. G., in: Colloid Science, ed. by H. R. Kruyt, Vol. 1, p. 226, Elsevier, Amsterdam 1952

[24] Schenkel, J. H., and J. A. Kitchener, Trans. Faraday Soc. **561**, 161 (1960)

[25] Gregory, J., J. Colloid Interface Sci. **83**, 138 (1981)

[26] Casimir, H. B., and D. Polder, Phys. Rev. **73**, 360 (1948)

[27] Clayfield, E. J., E. C. Lumb, and P. H. Mackey, J. Colloid Interface Sci. **37**, 382 (1971)

[28] Czarneckij, J., and V. Itšenskij, J. Colloid Interface Sci. **98**, 590 (1984)

[29] Czarneckij, J., and V. Itšenskij, Kolloidny Ž. (in press)

[30] Pashley, R. M., J. Colloid Interface Sci. **80**, 153 (1983)

[31] Vold, M. J., J. Colloid Sci. **16**, 1 (1961)

[32] Buske, N., H. Sonntag, and J. Götze, Colloid Surf. **12**, 195 (1984)

[33] Sonntag, H., and R. Kolesnikowa, Z. Phys. Chem. **261**, 226 (1980)

[34] Vincent, B., J. Colloid Interface Sci. **66**, 68 (1978)

[35] Hesselink, F. Th., J. Phys. Chem. **75**, 65 (1971)

[36] Hesselink, F. Th., J. Th. G. Overbeek, and A. Vrij, Z. Phys. Chem. **75**, 2094 (1971)
[37] Hesselink, F. Th., J. Polymer Sci. Symp. **61**, 439 (1977)
[38] Derjaguin, B. V., Kolloid Z. **69**, 155 (1934)
[39] Scheutjens, J. M., and G. J. Fleer, J. Phys. Chem. **83**, 1619 (1979)
[40] Scheutjens, J. M., and G. J. Fleer in: The Effect of Polymers on Dispersion Properties, ed. T. F. Tadros, Academic Press, London 1982
[41] Derjaguin, B. V., N. V. Čuraev, and V. M. Muller, Surface Forces, Nauka, Moscow 1985, English translation in preparation
[42] Derjaguin, B. V., and S. M. Sorin, Proc. IInd International Congress of Surface Active Substances 2, 145 (1957)
[43] Pashley, R. M., and J. A. Kitchener, J. Colloid Interface Sci. **71**, 491 (1979)
[44] Rabinovič, J. I., B. V. Derjaguin, and N. V. Čuraev, Adv. Colloid Interface Sci. **16**, 63 (1982)
[45] Israelachvili, J. N., Faraday Discuss. Chem. Soc. **65**, 20 (1978)
[46] Healy, T. W., A. Homola, and R. O. James, Faraday Discuss. Chem. Soc. **65**, 156 (1978)
[47] Ruckenstein, E., and D. Schiby, Chem. Phys. Lett. **95**, 439 (1983)
[48] Pashley, R. M., and J. P. Quirk, Colloid Surf. **9**, 1 (1984)
[49] Pashley, R. M., and J. N. Israelachvili, J. Colloid Interface Sci. **101**, 511 (1984); **102**, 23 (1984)

Chapter 2

[1] Einstein, A., Ann. Phys. (Leipzig) **17**, 549 (1905); **19**, 371 (1906)
[2] Von Smoluchowski, M., Ann. Phys. (Leipzig) **21**, 756 (1906)
[3] Langevin, D., C. R. Acad Sci. Paris **146**, 530 (1908)
[4] Von Smoluchowski, M., Phys. Z. **27**, 530 (1908)
[5] Von Smoluchowski, M., Z. Phys. Chem. **92**, 129 (1917)
[6] Derjaguin, B. V., Dokl. Akad. Nauk SSSR **109**, 967 (1956)
[7] Derjaguin, B. V., and V. M. Muller, Dokl. Akad. Nauk SSSR **176**, 869 (1967)
[8] Spielmann, L., J. Colloid Interface Sci. **33**, 562 (1970)
[9] Honig, E. P., G. J. Roebersen, and P. H. Wiersema, J. Colloid Interface Sci. **36**, 97 (1971)
[10] Brenner, H., Chem. Eng. Sci. **16**, 242 (1961)
[11] McGown, D. N. L., and G. D. Parfitt, J. Phys. Chem. **71**, 449 (1967)
[12] Fuchs, N. A., Z. Phys. **89**, 736 (1934)
[13] Reerink, H., and J. Th. G. Overbeek, Faraday Discuss Chem. Soc. **18**, 74 (1954)
[14] Ottewill, R. H., and J. N. Shaw, Faraday Discuss Chem. Soc. **42**, 154 (1966)
[15] Martynov, G. A., and V. M. Muller, Proc. IVth Conference on Surf. Forces in thin Films and dispersed systems 1969, Moscow 1972, p. 7
[16] Frens, G., and J. Th. G. Overbeek, J. Colloid Interface Sci. **38**, 376 (1972)
[17] Frens, G., Faraday Discuss. Chem. Soc. **65**, 146 (1978)
[18] Hogg, R., and K. C. Yang, J. Colloid Interface Sci. **56**, 573 (1976)
[19] Bagchi, P., Adv. Chem. Ser. 145 **(1975)**
[20] Müller, H., Kolloid Z. **38**, 1 (1926)
[21] Müller, H., Kolloidchem. Beih. **26**, 257 (1928)
[22] Müller, H., Kolloidchem. Beih. **27**, 223 (1928)
[23] Valioulis, J. A., and E. J. List, Adv. Colloid Interface Sci. **20**, 1 (1984)
[24] Cornell, R. M., J. W. Goodwin, and R. H. Ottewill, J. Colloid Interface Sci. **71**, 254 (1979)
[25] Marmur, A., J. Colloid Interface Sci. **72**, 41 (1979)
[26] Ruckenstein, E., J. Colloid Interface Sci. **66**, 531 (1970)
[27] Prieve, D., and E. Ruckenstein, J. Colloid Interface Sci. **73**, 539 (1980)
[28] O'Neill, M. E., Proc. Cambridge Philos. Soc. **66**, 407 (1969)
[29] Reynolds, P. A., and J. W. Goodwin, Colloid Surf. **11**, 145 (1984)
[30] Overbeek, J. Th. G., in Colloid Science, ed. by H. R. Kruyt, Vol. 1, p. 266, Elsevier, Amsterdam 1952

Chapter 3

[1] Von Smoluchowski, M., Phys. Z. **27**, 585 (1916)
[2] Von Smoluchowski, M., Z. Phys. Chem. **92**, 129 (1917)
[3] Müller, H., Kolloid Z. **38**, 1 (1926)
[4] Müller, H., Kolloid chem. Beih. **26**, 257 (1928)
[5] Mulholland, G. W., T. G. Lee, and H. R. Baum, J. Colloid Interface Sci. **62**, 406 (1977)
[6] Friedlander, S. K., and C. S. Wang, J. Colloid Interface Sci. **22**, 126 (1966)
[7] Wang, C. S., and S. K. Friedlander, J. Colloid Interface Sci. **24**, 170 (1967)
[8] Cohen, E. R., and E. U. Vaughan, J. Colloid Interface Sci. **35**, 612 (1971)
[9] Lee, K. W., J. Colloid Interface Sci. **92**, 315 (1983)
[10] Lindauer, G. C., and A. W. Castleman, Nucl. Sci. Eng. **43**, 212 (1971)
[11] Swift, D. L., and S. K. Friedlander, J. Colloid Sci. **19**, 621 (1964)
[12] Yoshida, T., K. Okuyama, Y. Kousaka, and Y. Y. Kita, Chem. Eng. Japan **8**, 317 (1975)
[13] Martynov, G. A., and V. M. Muller: Surface forces in thin films and dispersed systems, Moscow 1972, p. 7ff.
[14] Schilov, V., H. Sonntag, H. Gedan, and H. Lichtenfeld, Colloid Surf. Sci. 20, 303 (1986)
[15] Kruyt, H. R., Colloid Science, Elsevier, New York 1952, p. 285
[16] Landau, L. D., and E. M. Lifschitz, Lehrbuch der theoretischen Physik, Vol. 1, Akademie Verlag, Berlin 1980
[17] Thomas, I. L., K. M. McCorkle, J. Colloid Interface Sci. **36**, 110 (1971)
[18] Florek, Th., Thesis, AdW der DDR, Berlin 1981
[19] Florek, Th., H. Sonntag, and V. Schilov, Adv. Colloid Interface Sci. **16**, 337 (1982)
[20] Happel, J., and H. Brenner: Low Reynolds Number Hydrodynamics, Prentice Hall, Englewood Cliffs, New Jersey 1965
[21] Siedentopf, H., and R. Zsigmondy, Ann. Phys. (Leipzig) **10**, 1 (1903)
[22] Derjaguin, B. V., G. J. Vlasenko, Russian patent No. 86851 (1944)
[23] Derjaguin, B. V., and G. J. Vlasenko, Dokl. Akad. Nauk **63**, 155 (1948)
[24] Derjaguin, B. V., and G. J. Vlasenko, J. Colloid Sci. **17**, 605 (1962)
[25] Gucker, F. F., C. T. Okonski, H. B. Pickard, and G. N. Pitts, J. Am. Chem. Soc. **69**, 2422 (1947)
[26] Gucker, F. F., and C. T. Okonski, Chem. Rev. **44**, 373 (1949)
[27] Gucker, F. F., and C. T. Okonski, J. Colloid Sci. **4** 541 (1964)
[28] McFadayen, P. F., and A. Smith, J. Colloid Interface Sci. **45**, 573 (1973)
[29] Watillon, A., and F. van Grunderbeek, Bull. Soc. Chim. Belg. **63**, 115 (1954)
[30] Ottewill, R. H., and D. J. Wilkins, J. Colloid Interface Sci. **15**, 512 (1962)
[31] Buske, N., H. Gedan, H. Lichtenfeld, W. Katz, and H. Sonntag, Colloid Polym. Sci. **258**, 1303 (1980)
[32] Bartholdi, M., G. C. Salzmann, R. D. Hiebert, and M. Kerker, Appl. Opt. **19**, 1573 (1980)
[33] Cummins, P. G., E. J. Staples, L. G. Thomson, A. L. Smith, and L. Pope, J. Colloid Interface Sci. **92**, 189 (1983)
[34] Gedan, H., Thesis, AdW der DDR, Berlin 1984
[35] Walsh, D. J., J. Anderson, A. Parker, and M. J. Dix, Colloid Polym. Sci. **259**, 1003 (1981)
[36] Wyatt, P. J., and D. Philipps, J. Colloid Interface Sci. **39**, 125 (1972)
[37] Cooke, D. D., and M. Kerker, J. Colloid Interface Sci. **42**, 150 (1973)
[38] Higuchi, W. J., R. Okada, and A. P. Lemberger, J. Pharm. Sci. **51**, 683 (1962)
[39] Higuchi, W. J., R. Okada, G. A. Stelter, and A. P. Lemberger, J. Pharm. Sci. **52**, 49 (1963)
[40] Matthews, B. A., and C. T. Rhodes, J. Pharm. Sci. **57**, 558 (1967)
[41] Ottewill, R. H., and J. Shaw, Faraday Discuss. Chem. Soc. **42**, 154 (1966)
[42] Lichtenbelt, J. W. Th., H. J. M. Ras, and P. H. Wiersema, J. Colloid Interface Sci. **46**, 522 (1974)
[43] Heller, W., R. M. Tabibian, J. Colloid Sci. **12**, 25 (1957)
[44] Lichtenbelt, J. W. Th., C. Pathmamanoharan, and P. H. Wiersema, J. Colloid Interface Sci. **49**, 281 (1974)
[45] Melik, D. H., and H. S. Fogler, J. Colloid Interface Sci. **92**, 161 (1983)

[46] Pangonis, W. J., W. Heller, and A. Jacobson: Tables of Light Scattering Functions for Spherical Particles, Wayne State University Press Detroit 1957

[47] Gumprecht, R. O., and C. M. Sliepcevich, J. Phys. Chem. 57, 90 (1953)

[48] Bagchi, P., Adv. Chem. Ser. 145 (1975)

[49] Heller, W., Rev. Mod. Phys. 14, 390 (1942)

[50] Sokorov, S., T. Vorobewa, and S. Stoylov, J. Polym. Sci. 44, 142 (1974)

[51] Oaklay, D. M., and B. R. Jennings, J. Colloid Interface Sci. 91, 188 (1983)

[52] Stoylov, S. P., Adv. Colloid Interface Sci. 3, 45 (1971)

[53] Stoylov, S. P., V. N. Schilov, S. S. Duchin, S. Sokerov, and V. Petkanchin: Elektrooptica Kolloidov, Naukova Dumka, Kiev 1977

[54] Zsigmondy, R., Z. Phys. Chem. 92, 600 (1918)

[55] Westgren, A., and J. Reitstötter, Z. Phys. Chem. 92, 750 (1918)

[56] Tuorilla, A., Kolloidchem. Beih. 22, 193 (1926)

[57] Kruyt, H. R., and E. van Arkel, Rec. Trav. Chim. Pays-Bas 39, 656 (1920)

[58] Kruyt, H. R. and E. van Arkel, Rec. Trav. Chim. Pays-Bas 40, 169 (1921)

[59] Kudravzeva, N. M., and B. V. Derjaguin, Issledovanie f. obl. pov. sil. Isd. ANSSSR, Moscow 1961

[60] Derjaguin, B. V., and N. M. Kudravzeva, Kolloidny Ž. 26, 61 (1964)

[61] Watillon, A., M. Romerowski, and F. van Grunderbeek, Bull. Soc. Chim. Belg. 68, 450 (1959)

[62] Černoberežskij, M. T., E. V. Golikova, and L. V. Malinovskaja: Sb. Poverchnostij sil v tonkich plenkach i ustoitschivost kolloidov Izd. Nauka, Moscow 1974, p. 249

[63] Hatton, W. P., P. F. McFadayen, and A. L. Smith, J. Chem. Soc. Faraday Trans. I 70, 655 (1974)

[64] Kotera, A., K. Furusawa, and Y. Takeda, Kolloid Z. Z. Polym. 239, 667 (1970)

[65] Ottewill, R. H., and T. Walker, J. Chem. Soc. Faraday Trans. I 70, 917 (1974)

[66] Gedan, H., H. Lichtenfeld, H. Sonntag, and H. J. Krug, Colloid Surf. 11, 199 (1984)

[67] van der Scheer, M., A. Tanke, and C. Smolders, Faraday Discuss. Chem. Soc. 65, 264 (1978)

[68] Lips, A., C. Smart, and E. Willis, J. Chem. Soc. Trans. Faraday Soc. 67, 2979 (1971)

[69] Lips, A., and E. Willis, J. Chem. Soc. Faraday Trans. 69, 1226 (1973)

[70] Ottewill, R. H., and M. C. Rastogi, J. Chem. Soc. Trans. Faraday Soc. 56, 866 (1960)

[71] Johnson, G. A., S. M. A. Lecchini, E. G. Smith, J. Clifford, and B. A. Pethica, Faraday Discuss. 42, 120 (1966)

[72] van der Scheer, M., A. Tanke, and C. Smolders, Faraday Discuss. Chem. Soc. 65, 264 (1978)

[73] Frens, G., and J. Th. G. Overbeek, J. Colloid Interface Sci. 38, 376 (1972)

[74] Frens, G., Faraday Discuss. 65, 146 (1978)

[75] Honig, E. P., G. J. Roebersen, and P. H. Wiersema, J. Colloid Interface Sci. 36, 97 (1971)

[76] Reerink, H., and J. Th. G. Overbeek, Faraday Discuss. 18, 74 (1954)

[77] Kruyt, M. R., and M. A. M. Klompe, Kolloid Beih. 54, 507 (1936)

[78] Frens, G., and J. Th. G. Overbeek, J. Colloid Interface Sci. 36, 286 (1972)

[79] Benitz, R., and F. MacRitchie, J. Colloid Interface Sci. 40, 310 (1972)

[80] Long, J. A., O. W. J. Osmond, and B. Vincent, J. Colloid Interface Sci. 42, 545 (1973)

[81] Smith, A. L., and L. Thompson, J. Colloid Interface Sci. 77, 557 (1981)

[82] Hiemenz, P. C., and R. D. Vold, J. Colloid Interface Sci. 20, 635 (1965)

[83] Vincent, B., J. Colloid Interface Sci. 42, 270 (1973)

[84] Overbeek, J. Th. G., J. Colloid Interface Sci. 58, 408 (1977)

[85] Lyklema, J., Pure Appl. Chem. 53, 2199 (1981)

[86] Golikova, E. V., V. I. Kučuk, L. L. Molčanova, and Ju. M. Černoberešskij, Kolloidny Ž. 45, 864 (1983)

[87] Bowen, M. St., M. L. Broide, and R. J. Cohen, J. Colloid Interface Sci. 105, 605 (1985)

[88] Muller, V., Thesis, Leningrad 1985

[89] Rarity, J. G., and K. J. Randle, J. Chem. Soc. Faraday Trans. I 81, 285 (1985)

[90] Rarity, J. G., and K. J. Randle, Opt. Acta 31, 629 (1984)

[91] Eisenlauer, J., and D. Horn, Colloid Surf. 14, 121 (1985)

[92] Cowell, C., R. Li-In-On, and B. Vincent, J. Chem. Soc. Faraday Trans. I 74, 337 (1978)

[93] Cowell, C., and B. Vincent, J. Colloid Interface Sci. 87, 518 (1982)

[94] Cowell, C., and B. Vincent, J. Colloid Interface Sci. **95**, 573 (1983)
[95] Thompson, L., J. Chem. Soc. Faraday Trans. I **80**, 1673 (1984)
[96] Thompson, L., and D. Pryde, J. Chem. Soc. Faraday Trans. I **77**, 2405 (1981)
[97] Sonntag, H., V. Schilov, H. Lichtenfeld, H. Gedan, and C. Dürr, Colloid Surf. **20**, 303 (1986)
[98] Smellie, R. H. J., and V. K. Latter, J. Colloid Sci. **23**, 589 (1958)
[99] Healy, T. W., and V. K. Latter, J. Phys. Chem. **66**, 1835 (1962)
[100] Hogg, R., J. Colloid Interface Sci. **102**, 232 (1984)
[101] Cahill, J., P. G. Cummins, E. J. Staples, and L. Thompson, Colloid Surf. **18**, 189 (1986)

Chapter 4

[1] Heller, W., in: Polymer Colloids II, ed. by R. M. Fitch, Plenum Press, New York 1980, p. 153
[2] Walker, J., Sci. Am. **244**, 154 (1981)
[3] Walker, J., Sci. Am. **241**, 144 (1979)
[4] Walsh, D. J., J. Anderson, A. Parker, M. J. Dix, Colloid Polym. Sci. **259**, 100 (1981)
[5] Davidson, J. A., E. A. Collins, H. S. Haller, J. Polym. Sci., C **35**, 235 (1971)
[6] Buske, N., H. Gedan, H. Lichtenfeld, W. Katz, H. Sonntag, Colloid Polym. Sci. **258**, 1303 (1980)
[7] Gulari, E., E. Gulari, Y. Tsunashima, B. Chu, J. Chem. Phys. **70**, 3965 (1979)
[8] Brown, J. C., P. N. Pusey, J. W. Goodwin, R. M. Ottewill, J. Phys. A: Math. Gen. **8**, 646, 1433 (1975)
[9] Pusey, P. N., R. J. A. Tough, Adv. Colloid Interface Sci. **16**, 143 (1982)
[10] Ross, D. A., H. S. Dhadwai, R. R. Dyott, J. Colloid Interface Sci. **64**, 553 (1978)
[11] Dyott, R. B., Microwaves Opt. Acoust. **2**, 13 (1978)
[12] Manegold, E.: Kapillarsysteme, Vol. 1, Strassenbau, Chemie, Technik Verlag, Heidelberg 1955
[13] Sonntag, H., K. Strenge: Koagulation und Stabilität disperser Systeme, VEB Deutscher Verlag der Wissenschaften, Berlin 1970; in English: Coagulation and Stability of Disperse Systems, Halsted Press, New York 1972
[14] von Weimarn, P. P., Kolloid-Z. **2**, 76 (1907)
[15] Ostwald, W.: Die Welt der vernachlässigten Dimensionen, 12nd Ed., Verlag Th. Steinkopf, Dresden 1949
[16] Bargeman, D., E. J. van Emmerik, C. A. Smolders, Proc. Intern. Congr. Colloid and Surface Sci., Vol. 1, p. 439 Akadémiai kiadó, Budapest 1975
[17] Erdi, N. Z., M. M. Cruz, O. A. Battista, J. Colloid Interface Sci. **28**, 36 (1968)
[18] Battista, O. A., N. Z. Erdi, C. F. Ferraro, F. J. Karasimski, J. Appl. Polym. Sci. **11**, 481 (1967)
[19] Rehbinder, P. A., N. W. Michailov, Rheol. Acta **1**, 361 (1961)
[20] Ur'ev, N. B.: Vysokokoncentrirovanie dispersnych sistem, Chimija, Moscow 1980
[21] Hogg, R., T. W. Healy, D. W. Fuerstenau, Trans. Faraday Soc., **62**, 1638 (1966)
[22] Devereux, and de Bruyn: Interaction of parallel plane double layers, MIT Press, Cambridge, Massachusetts, 1963
[23] Bleier, A., E. Matijević, J. Colloid Interface Sci. **55**, 510 (1976)
[24] Princen, L. H., M. DeVena-Peplinski, J. Colloid Interface Sci. **19**, 786 (1964)
[25] Luckham, P. F., B. Vincent, Th. F. Tadros, Colloid Surface **6**, 83, 101, 119 (1983)
[26] Visser, J., Adv. Colloid Interface Sci. **15**, 157 (1981)
[27] Sonntag, H., N. Buske, Kolloid-Z. Z. Polym. **248**, 1016 (1971)
[28] Kihara, T., N. H. Honda, J. Phys. Soc. Japan **20**, 15 (1965)
[29] Tanford, Ch.: The Hydrophobic Effect, J. Wiley, New York 1973
[30] Cecil, R., Nature **214**, 369 (1967)
[31] Strenge, K., H. Sonntag, Trudy VII mežd. kongr. PAV, Moscow 1978, Vol. 2/II, B, 675
[32] Israelachvili, J., R. Pashley, Nature **300**, 341 (1982)
[33] Moelwyn-Hughes, E. A., Physical Chemistry, 2nd Ed., p. 306, Pergamon Press, Oxford 1964
[34] Geiseler, G., H. Seidel: Die Wasserstoffbrückenbindung, Akademie-Verlag, Berlin 1977

[35] Schalek, E., A. Szegvari, Kolloid-Z. **33**, 320 (1923)
[36] Usher, F. L., Proc. Roy. Soc. London Ser. **A 125**, 143 (1924)
[37] Wadrop, M. M., Science **214**, 1016 (1981)
[38] Cornell, R. M., J. W. Goodwin, R. H. Ottewill, J. Colloid Interface Sci. **71**, 254 (1979)
[39] Strenge, K., H. Sonntag, Proc. intern. Conf. Colloid and Surface Sci., Akadémiai Kiadó, Vol. 1, p. 397, Budapest 1975
[40] Thomas, J. L., K. H. McCorcle, J. Colloid Interface Sci. **36**, 110 (1971)
[41] Hoffmann, K., Kolloid-Z. **103**, 161 (1943)
[42] Rees, A. L. G., J. Phys. Chem. **55**, 1340 (1951)
[43] Sonntag, H., Th. Florek, V. Šilov, Adv. Colloid Interface Sci. **16** 337 (1982)
[44] Florek, Th., Thesis, Akademie der Wissenschaften der DDR, Berlin 1981
[45] Strenge, K., Diplomarbeit, Tech. Universität Dresden, 1961
[46] Hachisu, S., J. Colloid Interface Sci. **55**, 499 (1976)
[47] Rehbinder, P. A., I. Vlodavets: Ideen des exakten Wissens, 1, p. 5, Stuttgart 1971
[48] Vadas, E. B., H. C. Goldsmith, S. G. Mason, J. Colloid Interface Sci. **43**, 630 (1973)
[49] Žarkich, N. I., V. N. Šilov, Kolloidny Ž. **44**, 567, 571 (1982);
[50] Van Olphen, H.: An Introduction to Clay Colloid Chemistry, 2nd Ed., Wiley-Interscience, New York 1977
[51] Thiessen, P. A., Z. Elektrochem. **48**, 675 (1942)
[52] James, A. E., D. J. A. Williams, Adv. Colloid Interface Sci. **17**, 219 (1982)
[52a] James, A. E., D. J. A. Williams, J. Colloid Interface Sci. **55**, 79 (1978)
[53] Kuhn, A.: Kolloidchemisches Taschenbuch, Akad. Verlagsgesellschaft Geest & Portig, Leipzig 1960
[54] In: Sci. News **121**, 246 (1982)
[55] Okamoto, S., S. Hachisu, J. Colloid Interface Sci. **44**, 30 (1973)
[56] Van den Tempel, M., Rheol. Acta **12**, 115 (1958)
[57] Papenhuizen, J. M., Rheol. Acta **11**, 73 (1957)
[58] Strenge, K., H. Pilgrimm, Colloid Polym. Sci. **261**, 855 (1983)
[59] Mason, G., J. Colloid Interface Sci. **35**, 279 (1971)
[60] Medalia, A. I., Surf. Colloid Sci. **4**, 61 (1971)
[61] Vold, M. J., J. Colloid Sci. 14, 168 (1959)
[62] Vold, M. J., J. Phys. Chem. 64, 1616 (1960)
[63] Vold, M. J., J. Phys. Chem. 63, 1608 (1959)
[64] Vold, M. J., J. Colloid Sci. 18, 684 (1963)
[65] Hutchinson, H. P., D. N. Sutherland, Nature 206, 1036 (1965)
[66] Sutherland, D. N., J. Colloid Sci. **22**, 300 (1966)
[67] Sutherland, D. N., J. Colloid Sci. **25**, 373 (1967)
[68] Sutherland, D. N., J. Goodarz-Nia, Chem. Eng. Sci. 26, 2071 (1971)
[69] Goodarz-Nia, J., D. N. Sutherland, Chem. Eng. Sci. **30**, 402 (1975)
[70] Sutherland, D. N., Nature **226**, 1241 (1970)
[71] Deutch, J. M., I. Oppenheimer, J. Chem. Phys. **54**, 3542 (1971)
[72] Ermak, D. L., J. A. McLammon, J. Chem. Phys. **69**, 1352 (1978)
[73] Bacon, J., E. Dickinson, R. Parker, N. Anastasiou, M. Lal, J. Chem. Soc. Faraday Trans II, **79**, 91 (1983)
[74] Makarov, A. S., V. A. Suško, Kolloidny Ž. **51**, 795 (1979)
[75] Makarov, A. S., V. A. Suško, N. N. Kruglickij, Ukr. Chim. Ž. **45**, 438 (1979)
[76] Fisher, M. E., J. W. Essam, J. Math. Phys. **2**, 609 (1961)
[77] Domb, C., Adv. Phys. **9**, 283 (1960)
[78] Kirkpatrick, S., Rev. Mod. Phys. **45**, 579 (1973)
[79] Lagourette, B., J. Peyrelasse, C. Boned, M. Clausse, Nature, **281**, 60 (1979)
[80] Strenge, K., Dissertation B (Sc. D. thesis), Akademie der Wissenschaften der DDR, Berlin 1976
[81] Makarov, A. S., V. A. Suško, Visnik Akad. Nauk URSR **7**, 62 (1977)
[82] Aguf, I. A., T. N. Orkina, Kolloidny Ž. **41**, 403 (1979)
[83] Mewis, J., L. Hellinckx, Rheol. Acta **11**, 203 (1972)

[84] Helsen, J. A., R. Govaerts, G. Schoukens, J. De Graeuwe, J. Mewis, J. Phys. E, Sci. Instrum. **11**, 139 (1978)

[85] Bensley, C. N., R. J. Hunter, J. Colloid Interface Sci. **92**, 436 (1983)

[86] Efremov, I. F. Periodičeskie kolloidnye struktury, Chimija, Leningrad 1971

[87] Efremov, I. F.: Periodic colloid structures, Surf. Colloid Sci. **8**, 85 (1976)

[88] Hiltner, P. A., I. M. Krieger, J. Phys. Chem. **73**, 2386 (1969)

[89] Hiltner, P. A., Y. S. Papir, I. M. Krieger, J. Phys. Chem. **75**, 1881 (1971)

[90] Barnes, Ch. J., D. Y. C. Chan, D. H. Everett, D. E. Yates J. Chem Soc., Faraday Trans. II **74**, 138 (1982)

[91] Krieger, I. M., Adv. Colloid Interface Sci. **3**, 111 (1972)

[92] Fryling, C. F., J. Colloid Interface Sci., **18**, 713 (1963)

[93] Brodnyan, J. G., E. L. Kelley, J. Colloid Interface Sci. **19**, 488 (1964); and **20**, 7 (1965)

[94] Wang, Y., J. Colloid Interface Sci. **32**, 633 (1970)

[95] Hachisu, S., Y. Kabayashi, A. Kose, J. Colloid Interface Sci. **42**, 342 (1973)

[96] Ramsay, J. D. F., B. O. Booth, J. Chem. Soc. Faraday Trans. I **79**, 173 (1983)

[97] Saunders, J. N., Nature **204**, 1151 (1964)

[98] Saunders, J. N., Acta Crystallogr. A **24**, 427 (1968)

[99] Iler, R. K., Nature **207**, 472 (1965)

[100] Saunders, J. N., M. J. Murray, Nature **275**, 201 (1978)

[101] Hachisu, S., S. Yoshimura, Nature **283**, 188 (1980)

[102] Hachisu, S., K. Takano, Adv. Colloid Interface Sci. **16**, 233 (1982)

[103] Snook, I., W. van Megen, J. Chem. Soc. Faraday Trans. II **72**, 216 (1976)

[104] Snook, I., W. van Megen, Chem. Phys. Lett. **33**, 156 (1975)

[105] van Megen, W., I. Snook, J. Colloid Interface Sci. **53**, 172 (1975)

[106] van Megen, W., I. Snook, J. Colloid Interface Sci. **57**, 40, 47 (1976)

[107] van Megen, W., I. Snook, Chem. Phys. Lett. **35**, 399 (1975)

[108] van Megen, W., I. Snook, Faraday Disc. Chem. Soc. **65**, 92 (1978)

[109] Gaylor, K., W. van Megen, I. Snook, J. Chem. Soc., Faraday Trans. II **75**, 451 (1979)

[110] Snook, I., R. O. Watts, J. Colloid Interface Sci. **77**, 131 (1980)

[111] Keavey, R. P., P. Richmond, J. Chem. Soc., Faraday Trans. II **72**, 773 (1976)

[112] Benner, J. A., L. R. White, Colloid Surf. **3**, 371 (1981)

[113] Marcelja, S., D. J. Mitchell, B. W. Ninham, Chem. Phys. Lett. **43**, 353 (1976)

[114] Ise, N., T. Okubo, S. Sugimara, J. Chem. Phys. **78**, 563 (1983)

[115] Ise, N., T. Okubo, Acc. Chem. Res. **13**, 303 (1980)

[116] Mitchell, D. J., B. W. Ninham, Adv. Colloid Interface Sci. **9**, 57 (1978)

[117] Wiegner, E., Trans. Faraday Soc., **34**, 678 (1938)

[118] Brown, G. H., Am. Sci. **60**, 64 (1972)

[119] Brown, G. H., P. P. Crooker, Chem. Eng. News **61**, 24 (1983)

[120] Demus, D., L. Richter: Textures of Liquid Crystals, Verlag Chemie, Weinheim 1978

[121] Crandall, R. S., R. Williams, Science **198**, 293 (1977)

[122] Williams, R., R. S. Crandall, Phys. Lett. A **48**, 225 (1974)

[123] Clark, N. A., A. J. Hurd, B. J. Ackerson, Nature **281**, 57 (1979)

[124] Oshima, H., T. W. Healy, L. R. White, J. Colloid Interface Sci. **90**, 17 (1982)

[125] Morawetz, H.: Macromolecules in Solution, Interscience, New York 1965

[126] Muth, F., Kolloid-Z. **41**, 97 (1927)

[127] Arp, P. A., S. G. Mason, Colloid Polym. Sci. **255**, 1165 (1977)

[128] Vinogradov, G. V., Yu. F. Deinega, J. Inst. Petroleum **52**, 279 (1966)

[129] Ezernack, D., E. McLaughlin, J. Phys. E, Sci. Instr. **14**, 812 (1981)

[130] In: Isobretatel i racionalisator **11** (1979) (from the Lykov Inst. for Heat and Mass Exchange, Lab. of Rheophysics, Prof. S. P. Shulman, Belorusski Akad. Nauk)

[131] USSR Patent No. 546075

[132] USSR Inventor Papers No's 279767; 488294; 498699

[133] Manegold, E.: Kapillarsysteme, Vol. 1, Straßenbau, Chemie, Technik Verlag, Heidelberg 1955, p. 114

[134] Rosensweig, R. E., Sci. Am. **245**, 124 (1982)

[135] Papell, S. S., US Patent 3,215,572, Nov. **2**, 1965 (NASA)
[136] Bibik, E. E., Kolloidny Ž. **35**, 1141 (1973)
[137] Bibik, E. E., N. M. Gubanov, I. S. Lavrov, I. Ju. Mišuris, Leningradski Inst. im. Lensoveta, mežruzovskij sbornik trudov, No. 1, Leningrad 1976, p. 27
[138] Khalafalla, S. E., Chem.-Tech. **5**, 540 (1975)
[139] Buske, N., H. Sonntag, Z. Chem. **15**, 413 (1975)
[140] Buske, N, Freiberger Forschg.-Hefte A **664**, 215 (1980)
[141] Buske, N., H. Sonntag, G. Kelbg, Wiss. Fortschr. **29**, 90 (1979)
[142] Bibik, E. E., I. S. Lavrov, Kolloidny Ž. **27**, 652 (1965)
[143] Scholten, P. C., D. L. A. Tjaden, J. Colloid Interface Sci. **73**, 254 (1980)
[144] Popplewell, J., S. W. Charles, Phys. Bull. **30**, 474 (1979)
[145] Hall, W. F., S. N. Busenberg, J. Chem. Phys. **51**, 137 (1969)
[146] McTague, J. P., J. Chem. Phys. **51**, 133 (1969)
[147] Rosensweig, R. E., R. Kaiser, G. Miskolczy, J. Colloid Interface Sci. **29**, 680 (1969)
[148] Shulman, S. P., V. I. Kordomskij, S. A. Demchuk, Rheol. Acta **17**, 160 (1978)
[149] Henning, G., Funkschau **50**, 1162 (1978)
[150] In: Bild der Wiss. **11**, 23 (1974)
[151] Zocher, H., K. Jakobson, Kolloidchem. Beih. **28**, 167 (1929)
[152] Bergmann, P., P. Löw-Beer, H. Zocher, Z. Phys. Chem. (Leipzig) A **181**, 301 (1938)
[153] Furusawa, K., S. Hachisu, J. Colloid Interface Sci. **28**, 167 (1968)
[154] Maeda, Y. S. Hachisu, Colloid Surf. **6**, 1 (1983)
[155] Reiner, M., G. W. Scott Blair: Rheological Terminology, in: Rheology, ed. by F. R. Eirich, Vol. 4, p. 461, Academic Press, New York 1962
[156] Rehbinder, P. A., Pure Appl. Chem. **10**, 337 (1965)
[157] Strenge, K., Plaste Kautsch. **23**, 233 (1976)
[158] Einstein, A., Ann. Phys. (Leipzig) **19**, 289 (1906)
[159] Einstein, A., Ann. Phys. (Leipzig) **34**, 591 (1911)
[160] Arrhenius, S., Z. Phys. Chem. (Leipzig) **1**, 285 (1887)
[161] Happel, J., J. Appl. Phys. **28**, 1288 (1957)
[162] Rutgers, W. R., Rheol. Acta **2**, 305 (1962)
[163] Ostwald, W., Kolloid-Z. **36**, 99 (1925)
[164] Albers, W., J. Th. G. Overbeek, J. Colloid Sci., **15**, 489 (1960)
[165] Doroszkowski, A., R. Lambourne, J. Colloid Sci. **26**, 128 (1968)
[166] Cross, M. M., J. Colloid Sci. **20**, 417 (1965)
[167] Cross, M. M., J. Colloid Interface Sci. **33**, 30 (1970)
[168] Oldroyd, J. G., Proc. R. Soc. Ser. A **245**, 278 (1958)
[169] Hunter, R. J., S. K. Nicols, J. Colloid Interface Sci. **28**, 250 (1968)
[170] Bueche, F., J. Chem. Phys. **20**, 1959 (1952)
[171] Michaels, A. S., J. C. Bolger, Ind. Eng. Chem., Fund. **1**, 153 (1962)
[172] Firth, B. A., R. J. Hunter, J. Colloid Interface Sci. **57**, 248, 257, 266 (1976)
[173] van den Tempel, M., Adv. Colloid Interface Sci. **3**, 137 (1972)
[174] Rehbinder, P. A., N. A. Semenenko, Dokl. Akad. Nauk SSSR **64**, 835 (1949)
[175] Ludwik, P.: Die Kegelprobe, Springer Verlag, Berlin 1908
[176] Gorazdovskij, T. Ja., P. A. Rehbinder, Kolloidny Ž. **32**, 512 (1970)
[177] Kruglickij, N. N., A. S. Makarov, V. A. Suško, N. V. Polišžuk, Fis.-chim. Mech. (Kiev) **8**, 36 (1976)
[178] Strenge, K., Chem. Tech. **30**, 200 (1978)
[179] Kolbanovskaja, A. S., P. A. Rehbinder, Kolloidny Ž. **12**, 194 (1950)
[180] Švedov, F. N., Physique **8**, 341 (1889)
[181] Sonntag, H., K. Strenge, V. N. Schilov, Colloid Polym. Sci. **255**, 292 (1977)
[182] Sžukin E. D., P. A. Rehbinder, Acta Chim. Acad. Sci. Hungaria **76**, 281 (1973)
[183] Strenge, K., H. Sonntag, Colloid Polym. Sci. **252**, 133 (1974)
[184] Strenge, K., H. Sonntag, Colloid Polym. Sci. **260**, 638 (1982)
[184a] Strenge, K., H. Sonntag, Colloid Polym. Sci. **262**, 509 (1984)
[185] Visser, J., Adv. Colloid Interface Sci. **3**, 331 (1972)

[186] Svedberg, H. D. Pedersen: Die Ultrazentrifuge, Verlag Th. Steinkopf, Dresden, 1940
[187] Buscall, R., Colloid Surf. **5**, 269 (1982)
[188] Analytical ultracentrifuge, type 3170 B, MOM works, Budapest, Hungary
[189] Makarov, A. S., K. Strenge, V. A. Suško, H. Sonntag, N. N. Kruglickij, Dokl. Akad. Nauk SSSR **272**, 410 (1983)
[190] Meijer, A. E. J., W. J. van Megen, J. Lyklema, J. Colloid Interface Sci. **66**, 99 (1978)
[191] Barcley, L., A. Harrington, R. H. Ottewill, Kolloid-Z., Z. Polym. **250**, 655 (1972)
[192] Homola, A., A. A. Robertson, J. Colloid Interface Sci. **54**, 286 (1976)
[193] Dickinson, E., A. Patel, Colloid Polym. Sci. **257**, 431 (1979)
[194] Homola, A., J. Snook, W. van Megen, J. Colloid Interface Sci. **61**, 493 (1977)
[195] Feldkamp, J. R., D. Swartzendruber, I. Shainberg, Colloid Polym. Sci. **261**, 277 (1983)
[196] Bode, R., H. Ferch, H. Fratzscher, Kautsch., Gummi, Kunstst. **20**, 578 (1967)
[197] Tiller, F. M., Z. Khatib, J. Colloid Interface Sci. **100**, 55 (1984)
[198] Beresford-Smith, B., D. Y. C. Chan, D. J. Mitchell, J. Colloid Interface Sci. **105**, 216 (1985)
[199] Van Megen, W., I. Snook, Adv. Colloid Interface Sci. **21**, 119 (1984)
[200] Dickinson, E., J. Colloid Interface Sci. **98**, 587 (1983)
[201] Buske, N., H. Sonntag, Th. Götze, Colloid Surf. **12**, 195 (1984)
[202] Stangroom, J. E., Phys. Technol. **14**, 290 (1983)
[203] Unpublished results.
[204] Tomita, M., T. G. M. van de Ven, J. Phys. Chem. **89**, 1291 (1985)
[205] Žarkich, E., V. Šilov, Kolloidny Ž. **44**, 571 (1982)
[206] Stewart, R. F., D. Sutton, Chem. Ind. 373 (1984)
[207] Israelachvili, J., R. Pashley, J. Colloid Interface Sci. **98**, 500 (1984)

Subject Index